Power Quality in Microgrids Based on Distributed Generators

Power Quality in Microgrids Based on Distributed Generators

Special Issue Editors

Ambrish Chandra
Hua Geng

MDPI • Basel • Beijing • Wuhan • Barcelona • Belgrade

MDPI

Special Issue Editors
Ambrish Chandra
Department of Electrical
Engineering, École de
Technologie Supérieure
Canada

Hua Geng
Tsinghua University
China

Editorial Office
MDPI
St. Alban-Anlage 66
4052 Basel, Switzerland

This is a reprint of articles from the Special Issue published online in the open access journal *Energies* (ISSN 1996-1073) in 2018 (available at: https://www.mdpi.com/journal/energies/special_issues/ Power_Quality_in_Microgrids_Based_on_Distributed_Generators).

For citation purposes, cite each article independently as indicated on the article page online and as indicated below:

LastName, A.A.; LastName, B.B.; LastName, C.C. Article Title. *Journal Name* **Year**, *Article Number, Page Range*.

ISBN 978-3-03928-006-3 (Pbk)
ISBN 978-3-03928-007-0 (PDF)

Contents

About the Special Issue Editors

Ambrish Chandra (Professor) has served as Full Professor of Electrical Engineering at École de technologie supérieure (ÉTS), Montréal, since 1999. He received his B.E. degree from the University of Roorkee (presently IITR), India, in 1997; M. Tech. from IIT Delhi in 1980; and Ph.D. from University of Calgary in 1987. Before joining as Associate Professor at ÉTS in 1994, he was a faculty member of IITR. From 2012 to 2015, he was Director of the multidisciplinary graduate program on Renewable Energy and Energy Efficiency at ÉTS. Presently, he is Director of the master program in Electrical Engineering at ETS. The primary focus of his work is related to the advancement of new theory and control algorithms for power electronic converters for power quality improvement in distribution systems and integration of renewable energy sources. The key differentiator of his work is in its simplicity and the practicality of new solutions which have had significant impact. His research work is referred to by both engineers as well as researchers around the world. His total Google citations exceed 13,500 and he has an h-index of 48 and i10-index of 156. He is coauthor of the John Wiley book "Power Quality—Problems and Mitigation Techniques". He is Fellow of many organizations, including IEEE, Canadian Academy of Engineering, Institute of Engineering and Technology U.K., and Engineering Institute of Canada, and registered as a Professional Engineer in Quebec. He is a Distinguished/Prominent Lecturer of the IEEE Power and Energy Society and the IEEE Industry Application Society. He was the recipient of the IEEE Canada P. Ziogas Electric Power Award in 2018.

Hua Geng (Research Professor) received his Ph.D. degree in control theory and application from Tsinghua University, Beijing, China, in 2008. From 2008 to 2010, he was a Postdoctoral Research Fellow at Ryerson University, Canada. He joined the Automation Department of Tsinghua University in June 2010 and now is a tenured Associate Professor. His current research interests include renewable energy conversion systems, flexible AC transmission systems (FACTS), microgrids, and digital control of power electronics. He has been the principle investigate (PI) or Co-PI for over 20 projects, including a National Excellent Young Scholar, Key Program, and China–UK joint Program of National Science Foundation of China (NSFC), and National Key R&D Program of China. Dr. Geng has published more than 150 technical papers, authored a China Machine Press book, and holds 10 issued Chinese patents. He was granted "Delta Young Scholar Award" by Delta Environmental and Educational Foundation in 2014 and "Young Professional Award" by China Power Supply Society (CPSS) in 2013. Dr. Geng is an IET Fellow, IEEE Senior Member, Chair of the Youth Working Committee of CPSS, Standing Director of CPSS. He serves as Editor of IEEE Trans. on Energy Conversion, IEEE Trans. on Sustainable Energy, and IEEE Power Engineering Letters, Associate Editor of IEEE Trans. on Industry Applications, Control Engineering Practice, Chinese Journal of Electrical Engineering, and CPSS Trans. on Power Electronics and Applications.

Preface to "Power Quality in Microgrids Based on Distributed Generators"

Microgrids are considered as a promising means for integrating various distributed generators (DGs) and loads resulting in benefits to the reliability, loss reduction, and carbon emission reduction, to name but a few. Power quality is an emerging issue in microgrids, and the problem is more complicated than that in conventional distribution systems because of the intermittent nature of renewable energy sources, as well as the increased infiltration o f n onlinear l oads a nd power electronic-interfaced DG systems. To address this important topic, Prof. Ambrish Chandra of ÉTS and Dr. Hua Geng in Tsinghua—both well-known experts in this field—have t aken t he lead in organizing this timely Special Issue on "Power Quality in Microgrids Based on Distributed Generators". The papers reported in this Special Issue reflect t he l atest r esearch r esults a nd new trends in the study of microgrid and related power quality issues, and can be used as a reference for students and researchers in this field.

Ambrish Chandra, Hua Geng
Special Issue Editors

energies

MDPI

Article

Time-Domain Voltage Sag State Estimation Based on the Unscented Kalman Filter for Power Systems with Nonlinear Components

Rafael Cisneros-Magaña [1], Aurelio Medina [1,*] and Olimpo Anaya-Lara [2]

[1] División de Estudios de Posgrado, Facultad de Ingeniería Eléctrica, Universidad Michoacana de San Nicolás de Hidalgo, Av. Francisco J. Múgica S/N, Morelia, Michoacán 58030, Mexico; rcisneros@dep.fie.umich.mx
[2] Institute for Energy and Environment, Department of Electronic and Electrical Engineering, University of Strathclyde, 204 George Street, Glasgow G1 1RX, UK; olimpo.anaya-lara@strath.ac.uk
* Correspondence: amedinr@gmail.com; Tel.: +52-443-327-9728

Received: 1 May 2018; Accepted: 23 May 2018; Published: 1 June 2018

Abstract: This paper proposes a time-domain methodology based on the unscented Kalman filter to estimate voltage sags and their characteristics, such as magnitude and duration in power systems represented by nonlinear models. Partial and noisy measurements from the electrical network with nonlinear loads, used as data, are assumed. The characteristics of voltage sags can be calculated in a discrete form with the unscented Kalman filter to estimate all the busbar voltages; being possible to determine the rms voltage magnitude and the voltage sag starting and ending time, respectively. Voltage sag state estimation results can be used to obtain the power quality indices for monitored and unmonitored busbars in the power grid and to design adequate mitigating techniques. The proposed methodology is successfully validated against the results obtained with the time-domain system simulation for the power system with nonlinear components, being the normalized root mean square error less than 3%.

Keywords: nonlinear dynamic system; power quality; power system simulation; state estimation; unscented Kalman filter; voltage fluctuation

---

1. Introduction

Power quality (PQ) is an important operation issue in any power system. Utilities must comply with strict standards, relating primarily harmonics, transients and voltage sags [1–4]. PQ depends on the power supply, the transmission and distribution systems and the electrical load condition. Voltage sags are among the adverse PQ effects; they can cause malfunction of electronic loads, and can reset voltage-sensitive loads [5,6]. The voltage sags characteristics in magnitude and duration are necessary to determine their effect in the grid and its loads. They constitute the majority of PQ problems, representing about 60% of them [7,8]. Among the problems that the nonlinear electrical components introduce to the power grid is the increase of harmonic distortion, which is an important effect to mitigate. Voltage sags have increased due to the use of nonlinear varying loads such as power electronic devices, smelters, arc furnaces and electric welders, the starting of large electrical loads, switching transients, connection of transformers and transmission lines, network faults, lightning strikes, network switching operations, among others [9].

The Kalman filter (KF) and the least squares methods have been used to estimate the voltage fluctuations in linear power systems [10–13]. PQ state estimation based on the KF uses a linear model, partial and noisy measurements from the system. In [14] the number of sags is estimated using a limited number of monitored busbars, recording the number of voltage sags during a determined period.

This research work proposes as an innovation, an alternative methodology based on the unscented Kalman filter (UKF) to perform the voltage sags state estimation (VSSE) in nonlinear load power networks; this method can also be applied to nonlinear micro grids. The VSSE determines the magnitude, duration and beginning-ending time of sags, with an observable system condition for the busbars voltages using the available measurements.

The KF has been applied to estimate harmonics and voltage transients in a signal [15], KF gain can be modified during the state estimation to reduce the estimation error [16], both references assess linear cases; [17] has proposed the UKF to detect sags in a voltage waveform. In this work, the UKF is extended to the nonlinear case to solve the time-domain VSSE, to estimate voltage sags in all busbars of a power system including nonlinear components. The UKF makes use of a power grid nonlinear model and noisy measurements from the same electrical network to estimate all the busbar voltages.

The extended Kalman filter (EKF) can be also applied to solve the nonlinear state estimation. The UKF error is slightly smaller when compared to the EKF error. This state estimation error increases in the filters when sudden variations are present, both being of about the same accuracy. The EKF can lead to divergence more easily than UKF, which shows good numerical stability properties.

The state estimation receives measurements from the power network, through a wide area measurement system (WAMS) and estimates the state vector, using algorithms such as the UKF. Practical implementation of the time-domain state estimation can be achieved with measuring instruments and data acquisition cards, capable of recording the voltage and current waveforms synchronously during several cycles, e.g., using the global positioning system (GPS) to time stamp the measurements [18–21]. The use of adequate communication channels like especially dedicated optical fibre links, allows to the measurements be sent to the control centre with high data updating rate, where they are received and numerically processed using computational systems with sufficient memory and adequate capability [9].

Due to economic reasons measurement technology for VSSE is currently limited, making the system underdetermined. The VSSE presents different problems from those of the traditional power system state estimation, where redundancy of measurements is possible [22].

The VSSE has been assessed in the frequency domain [14,23]. In this work, the UKF is proposed as an alternative method to obtain the time-domain VSSE. This approach makes possible the use of nonlinear models to represent more accurately the power system components and to obtain the results with a low state estimation error. The state estimation obtains the global or total system state that can be used to take corrective actions to mitigate the adverse effects of voltage sags, such as the network configuration change or control of flexible alternating current transmission system (FACTS) devices, e.g., the static synchronous compensator (STATCOM).

The time-domain UKF state estimation methodology can be used not only to estimate voltage sags but also to estimate over voltages, over currents or electromagnetic transients. The main objective of this work is to apply the UKF to obtain the VSSE, by addressing the dynamics of the nonlinear electrical networks and by estimating and delimiting the voltage sags in the time-domain. The case studies address short circuit faults and transient load conditions. The results are validated against the actual time-domain response of the power grid.

2. Dynamic State Estimation

The network model can be a set of first order differential equations to describe the dynamic state performance. The dynamic estimation data are the grid model with its inputs and a measurement set of selected outputs from the system during a determined number of cycles to define the measurement equation.

The KF dynamically follows the variations in the states, i.e., currents and voltages, detecting changes in the voltage waveform within less than half of a cycle and it is a good tool for instantaneous tracking and detection of voltage sags [24,25].

The KF solves the dynamic estimation, due to its recursive process [26,27]; being applied in linear cases. The UKF solves the dynamic estimation in nonlinear cases. In this work, the UKF estimates the nonlinear power system state under transient conditions, e.g., voltage sags [28]. Figure 1 describes the proposed VSSE methodology. The main steps are the nonlinear power system modelling and simulation, then UKF is applied to obtain the time-domain VSSE, and lastly the assessment of rms busbar voltages.

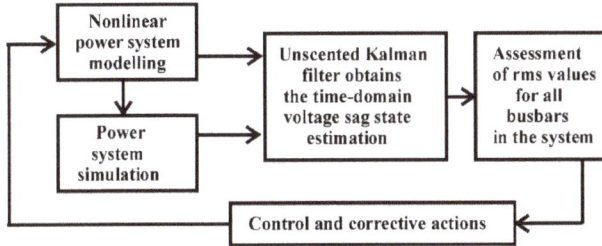

Figure 1. Time-domain UKF VSSE.

The UKF applies a deterministic sampling technique; i.e., the unscented transform (UT), which takes a set of sigma points near of their mean value. These points are propagated through the nonlinear model by evaluating the estimated mean and covariance [25]. The mean and covariance are encoded in the set of sigma points; these points are treated as elements of a discrete probability distribution, which has mean and covariance equal to those originally given. The distribution is propagated by applying the non-linear function to each point. The mean and the covariance of the transformed points represent the transformed estimate.

The main advantage of the UKF is the derivative free nonlinear state estimation, thus avoiding analytical or numerical derivatives [29,30]. The UT avoids the need of linearization using the Jacobian matrix as in the EKF, and it can be applied to any function, independently if it is differentiable or not. The UKF includes a Cholesky decomposition with an inverse matrix to evaluate the sigma points at each time step.

Inaccuracies of the model and its parameters can be taken into account with a statistical term w, called noise process. It accounts for the existence of phenomena such as the thermal noise of the electrical elements and the ambiguity in the accuracy of the parameters. Metering devices have errors and noise; they are represented by a statistical term v. In most cases, w and v have a Gaussian distribution. UKF is able to operate with partial, noisy, and inaccurate measurements [31,32].

3. Unscented Kalman Filter Methodology

The UT is based on the mean and covariance propagation by a nonlinear transform. The system and measurement nonlinear models can be represented as:

$$dx/dt = f(x, u, w) \tag{1}$$

$$y = h(x, u, v) \tag{2}$$

where $x \in \mathbb{R}^{n \times 1}$ is the state vector, u the known input vector of variable order, y the variable order output vector, f a nonlinear state function and h is a nonlinear output function, with n states and m measurements.

UKF uses a deterministic approach for mean and covariance calculation; $2n + 1$ sigma points are defined by using a square root decomposition of prior covariance. Sigma points propagation through the model (1) obtains the weighted mean and covariance. W_i represents the scalar weights, defined as:

$$W_0^{(m)} = \lambda/(n+\lambda) \tag{3}$$

$$W_0^{(c)} = \lambda/(n+\lambda) + \left(1 + \alpha^2 + \beta\right) \tag{4}$$

$$W_i^{(m)} = W_i^{(c)} = 1/(2(n+\lambda)), \qquad i = 1, \ldots, 2n \tag{5}$$

$$\lambda = \alpha^2(n+\kappa) - n \tag{6}$$

$$\gamma = \sqrt{n+\lambda} \tag{7}$$

where λ and γ are scaling parameters, α and κ determine the spread of sigma points; β is associated with the distribution of x. If Gaussian $\beta = 2$ is optimal, $\alpha = 10^{-3}$ and $\kappa = 0$ are normal values [30].

UT takes the sigma points with their mean and covariance values, and transform them by applying the nonlinear function f, and then the mean and covariance can be calculated for the transformed points. A weight W_i is assigned to each point.

UKF defines the n-state discrete-time nonlinear system from (1) and (2) as:

$$x_{k+1} = f(x_k, u_k, w_k, t_k) \tag{8}$$

$$y_k = h(x_k, u_k, v_k, t_k) \tag{9}$$

$$w_k \sim N(0, Q_k) \tag{10}$$

$$v_k \sim N(0, R_k) \tag{11}$$

Process noise w and measurement noise v are assumed stationary, zero-averaged and uncorrelated, $Q \in \mathbb{R}^{n \times n}$ and $R \in \mathbb{R}^{m \times m}$ are the covariance matrices for noises w and v, respectively.

UKF applies the following steps:

(a) Initialization, $k = 0$.

$$\hat{x}_0^+ = E(x_0) \tag{12}$$

$$P_0^+ = E\left[(x_0 - \hat{x}_0^+)(x_0 - \hat{x}_0^+)^T\right] \tag{13}$$

E is the expected value, P is the error covariance matrix, $+$ indicates update estimate or a posteriori estimate and $-$ project estimate or a priori estimate. Subscripts k and $k-1$ denote time instants $t_k = k\Delta t$ and $t_{k-1} = (k-1)\Delta t$, respectively, Δt is the time step.

(b) Sigma points assessment in matrix form by columns:

$$\chi_{k-1} = \left[\hat{x}_{k-1} \; \hat{x}_{k-1} + \gamma\sqrt{P_{k-1}}\hat{x}_{k-1} \; -\gamma\sqrt{P_{k-1}}\right] \tag{14}$$

(c) Update time step k from $k-1$.

$$\chi_{k|k-1}^* = f[\chi_{k-1}, u_{k-1}] \tag{15}$$

$$\hat{x}_k^- = \sum_{i=0}^{2n} W_i^{(m)} \chi_{i,\,k|k-1}^* \tag{16}$$

$$P_k^- = \sum_{i=0}^{2n} W_i^{(c)} \left[\chi_{i,\,k|k-1}^* - \hat{x}_k^-\right]\left[\chi_{i,\,k|k-1}^* - \hat{x}_k^-\right]^T + Q_k \tag{17}$$

$$\chi_{k|k-1} = \left[\hat{x}_k^- \; \hat{x}_k^- + \gamma \sqrt{P_k^-} \, \hat{x}_k^- - \gamma \sqrt{P_k^-} \right] \tag{18}$$

$$y_{k|k-1}^* = h\left[\chi_{k|k-1} \right] \tag{19}$$

$$\hat{y}_k^- = \sum_{i=0}^{2n} W_i^{(m)} y_{i,\, k|k-1}^* \tag{20}$$

χ matrix represents the sigma points; χ^* matrix represents the updated sigma points and y^* the updated output vector with sigma points.

(d)   Evaluate the error covariance matrices as:

$$P_{\hat{y}_k \hat{y}_k} = \sum_{i=0}^{2n} W_i^{(c)} \left[y_{i,\, k|k-1}^* - \hat{y}_k^- \right] \left[y_{i,\, k|k-1}^* - \hat{y}_k^- \right]^T + R_k \tag{21}$$

$$P_{x_k y_k} = \sum_{i=0}^{2n} W_i^{(c)} \left[\chi_{i,\, k|k-1}^* - \hat{x}_k^- \right] \left[y_{i,\, k|k-1}^* - \hat{y}_k^- \right]^T \tag{22}$$

(e)   UKF algorithm evaluates the filter gain K_k and updates the estimated state and the error covariance matrix.

$$K_k = P_{x_k y_k} P_{\hat{y}_k \hat{y}_k}^{-1} \tag{23}$$

$$\hat{x}_k^+ = \hat{x}_k^- + K_k \left(y_k - \hat{y}_k^- \right) \tag{24}$$

$$P_k^+ = P_k^- + K_k P_{\hat{y}_k \hat{y}_k} K_k^T \tag{25}$$

The steps (b)–(d), Equations (14)–(22), define the prediction stage, and the last step (e), Equations (23)–(25), defines the update stage, as in the KF algorithm [33,34]. The main objective of this work is to use the UKF formulation to estimate the busbar voltage waveforms, mainly at unmonitored busbars in the presence of voltage sags generated by faults and load transients.

Waveforms can be contaminated with noise, and the assumption of constant values for Q and R is valid when the noise characteristics are constant, like its standard deviation and variance. If the noise is varying, Q and R should be computed at each time step and an adaptive KF is a requirement [16]. UKF algorithm tracks the time-varying model and noise through the on-line calculation of Q and R. In this work, Q and R matrices are assumed constant, in order to mainly analyse the UKF application to time-domain VSSE.

UKF identifies the interval where the sags are present, as well as their magnitude, with an acceptable precision. By increasing the number of cycles, the UKF can identify the voltage characteristics during fault transient periods.

The number of points per cycle is of important concern to evaluate the time-domain state estimation with periodic signals. This number defines the sampling rate for the monitored signals. The sampled waveform is a sequence of values taken at defined time intervals and represents the measured variable. Interpolation can be used to adjust the number of points per cycle, linearly or nonlinearly [35]. In addition, the interpolation should be used carefully with discrete signals to satisfy the sampling theorem. The sampling rate defines the speed at which the input channels are sampled; this rate is defined in samples per cycle. To detect transients, high sampling rates compared with the fundamental frequency may be necessary [36].

Rms Value of Discrete Waveforms and Normalized Root Mean Square Error

The rms voltage magnitude can be determined by processing the discrete values for the voltage waveform according to the used data window size and the sampling frequency. The rms voltage magnitude V_{rms} for a discrete voltage signal can be calculated as:

$$V_{rms}(iN) = \sqrt{\left(\frac{1}{N} \sum_{j=(i-1)N+1}^{iN} V_j^2\right)} \quad i \geq 1 \tag{26}$$

where V_j is the sample voltage j and N is the number of samples per cycle taken in the sampling window; i is the sampled cycle. This expression can be applied to discrete voltage and current waveforms [22].

Normalized root mean square error (NRMSE) is used to validate the UKF-VSSE methodology; this error evaluates the state estimation residual between actually observed values and the estimated values; lower residual indicates less state estimation error. NRMSE is defined as:

$$\text{NRMSE} = \sqrt{\sum_{t=1}^{np} \frac{(\hat{y}_t - y_t)^2}{np}} / (y_{max} - y_{min}) \tag{27}$$

$\hat{y}$ is the estimated vector, y is the real or actually observed vector and np the number of elements of these vectors.

4. Case Studies

Figure 2 shows the modified IEEE 30-bus test system used in the case studies described next, assuming a three-phase base power of 100 MVA and a phase-to-phase base voltage of 230 kV. Lines 1–2, 1–4, 2–4, 2–5, 2–6, 4–6 and 5–6 are represented by an equivalent pi model and by series impedance the rest of lines; transformers 6–10, 4–12–13, 6–10–11, are represented by an inductive reactance, according to the IEEE 30-bus test power system [37].

The system is modified adding three nonlinear electrical loads, i.e., an electric arc furnace (EAF) to busbar 2, a nonlinear inductance to busbar 5 and a thyristor-controlled reactor (TCR) to busbar 6. The addition of these nonlinear elements gives the nonlinearity of (1) and (2). Appendix A gives additional parameters of nonlinear loads. Appendix B presents the nonlinear load models and their differential equations.

Generators are modelled as voltage sources connected to busbars through a series inductance. Linear electric loads are represented as constant impedances. Busbar voltages, line and load currents are defined as state variables to obtain the state space model for the power network; the measurements are function of these state variables.

The measurement locations are selected so that the busbar voltages are observable. Tables 1 and 2 show x and z vectors, respectively, to form the measurement equation by obtaining 103 measurements to estimate 110 state variables ($n = 110$, $m = 103$). The observation equation with this set of measurements has an underdetermined condition, but all the busbar voltages are observable to estimate the voltage sags. When busbar voltages are assessed and estimated other variables can be calculated, i.e., line currents or the TCR current.

The EAF real power and the nonlinear inductance current are included as nonlinear functions in the measurement equation ($z = Hx$) represented in the formulation by (2).

In the measurement matrix $H \in \mathbb{R}^{m \times n}$, each measurement is associated with its corresponding state variable (Table 2). The sampling frequency is at least 30.72 kHz, to obtain 512 samples per cycle, for a fundamental frequency of 60 Hz [24].

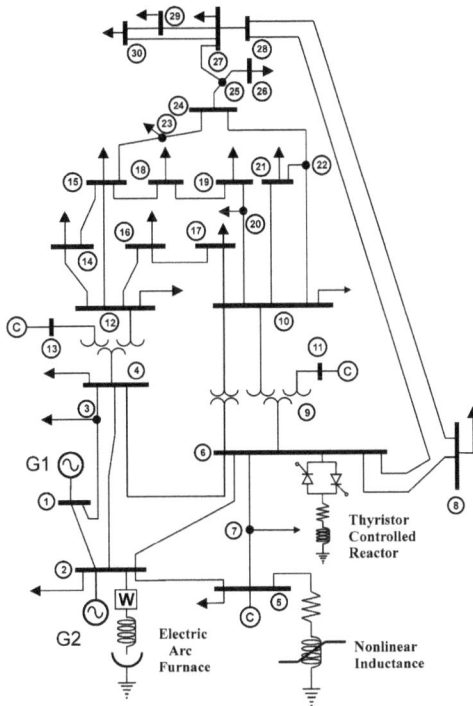

Figure 2. Modified IEEE 30-bus test power system with nonlinear loads at busbars 2, 5, and 6.

Table 1. State variable vector x.

Description	State Variable
Line currents	1–41
Busbar voltages	42–71
Generator currents	72–77
Busbar load currents	78–106
Nonlinear inductor magnetic flux	107
EAF current and arc radius	108–109
TCR current	110

Table 2. Measurements vector z.

Description	Output Variable
Line currents	1–38
Busbar voltages	42–68
Generator currents	72–77
Busbar load currents	78–106
Nonlinear inductor current	107
EAF real power	108–109

The conventional trapezoidal rule is used to solve the 110 first order ordinary differential equations set. To represent the power system, busbar voltages, line and load currents are defined as state variables; a step size of 512 points per period is used, i.e., 32.5 microseconds. The simulation time is set to 0.4 s or 24 cycles. The measurements are taken from this simulation and then are contaminated using randomly generated noise.

4.1. Case Study: UKF VSSE Short-Circuit Fault at Busbar 4

A transient condition is simulated by applying a single-phase to ground fault at busbar 4. The fault impedance is of 0.1 pu, to simulate a short-circuit fault, starting in cycle 13 (0.216 s) and ending in cycle 17 (0.283 s). This fault generates busbar voltage sags and swells, which can be estimated with the power network model, partial and noisy measurements from the system, and the UKF algorithm. The criterion to select this case study is to represent a transient fault in the transmission system and verify the proposed VSSE method.

Measurement noise is assumed with a signal to noise ratio (SNR) of 0.025 pu or 2.5%; while a SNR of 0.001 pu or 0.1% is assumed for the noise process. Figure 3 shows the busbar voltages 1–30, where the actual, the proposed UKF estimate and the difference between instantaneous values during the fault at busbar 4 are shown, corresponding to state variables 42–71.

The largest estimation error is present when the fault condition is removed at 0.283 s; this error is due to sudden changes in the busbar voltages. It is approximately 7%, but quickly decreases in the next three cycles to 1%. These voltage fluctuations are due to the short-circuit transient condition at busbar 4.

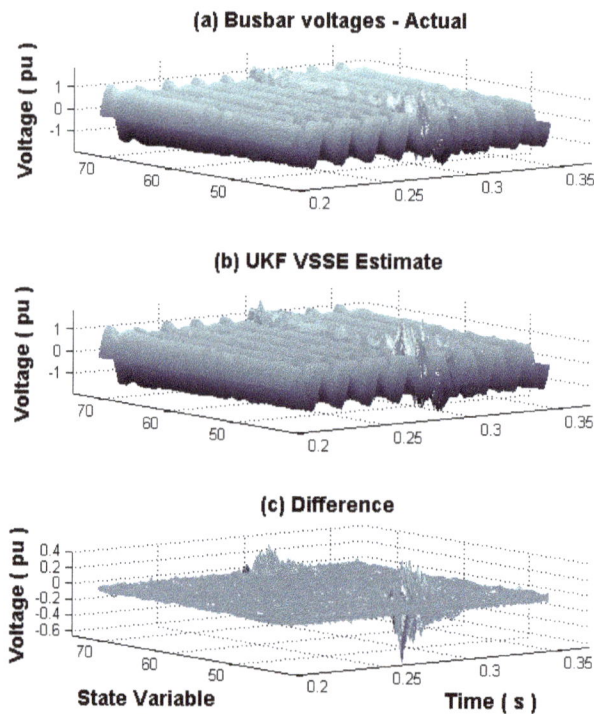

Figure 3. Busbar voltages (**a**) Actual, (**b**) UKF VSSE, (**c**) Difference, short-circuit at busbar 4 from 0.216 to 0.283 s.

Voltage waveforms for the faulted busbar 4 and for busbar 6, near to fault, are shown in Figure 4. Actual, UKF estimation and residual waveforms are illustrated. The presence of a voltage sag/swell condition at these busbars can be observed. Voltage sag lasts 4 cycles, while the fault condition is present, originating a reduction in the voltage magnitude of 12% for busbar 4 and 8% for busbar 6. Post-fault period begins at cycle 18, when the short circuit fault is removed. A voltage swell condition is present with a duration of two cycles and then the voltage eventually reaches the steady state. Residuals take considerable values during the voltage swell condition, the first two cycles of the post-fault period, and are due to the fast fluctuations of the state variables.

In Figure 4, NRMSE has been calculated using (27) to evaluate the state estimation error between actual and UKF estimated waveforms for the voltage busbars 4, and 6, during the 24 cycles under analysis, resulting on 2.5% and 1.2%, respectively.

Busbars 3–30 show a similar behavior as for busbars 4 and 6 during and after the short-circuit fault. The busbar voltage magnitude reduction mainly depends on the network topology, the load condition and the line impedance between the busbars. Fluctuations in the voltage waveforms at busbars are due to noisy measurements, network modelling, and the short-circuit fault used for the voltage sag/swell transient state estimation.

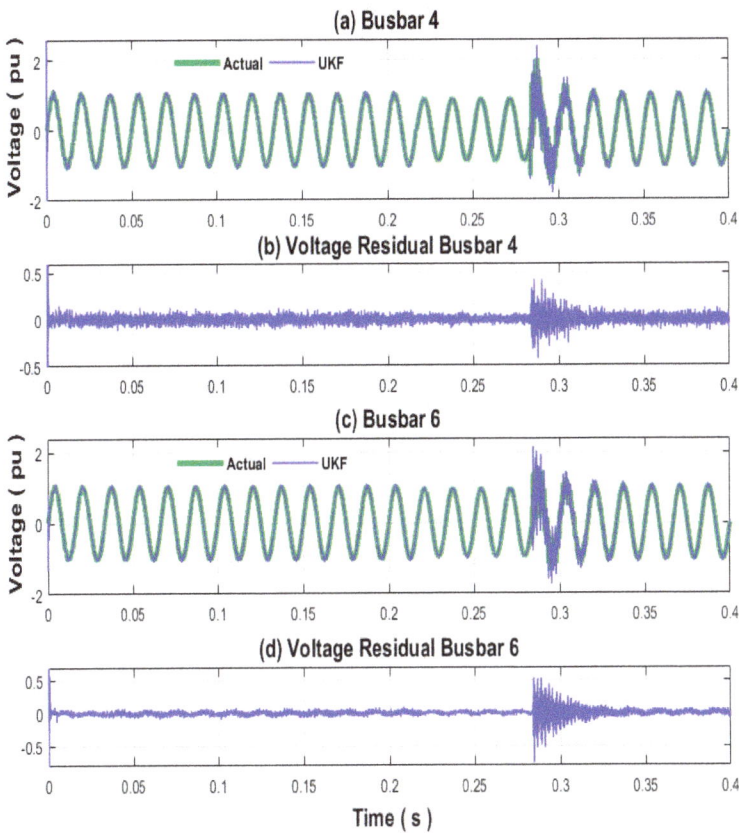

Figure 4. Actual, UKF estimation, and residuals of busbars 4 and 6, voltage sag 14–17 cycles from 0.216 to 0.283 s, voltage swell 18–19 cycles, short-circuit at busbar 4.

Line currents are shown in Figure 5 for the actual, UKF estimate and difference, respectively; with the fault condition at busbar 4 from 0.216 to 0.283 s.

The distribution of line currents in the power system is shown for the interval of study. This distribution represents the fault currents from generators to the faulted busbar 4, which can be observed in Figure 5 by the current fluctuations in the first state variables during and after the fault period. During the first cycle after fault clearance, the error increases to 12%, but once, this cycle ends the error decreases to around 1% in the post-fault period. The difference graph (c) presents this error at 0.283 s for the state variables representing the currents from generators to the faulted busbar 4.

Actual, UKF estimated currents and residuals of nonlinear components are illustrated in Figure 6, for the nonlinear inductance (a,b), the EAF (c,d) and the TCR (e,f).

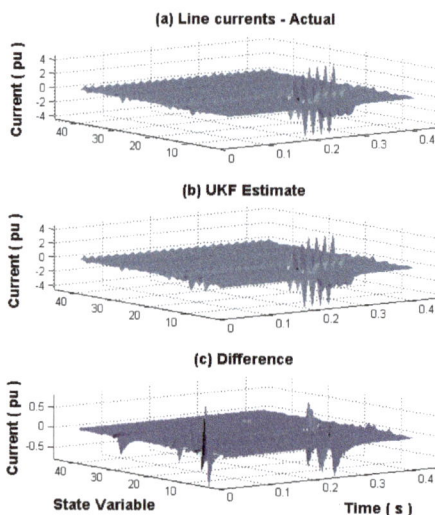

Figure 5. Line currents (**a**) Actual, (**b**) UKF estimation, (**c**) Difference, short-circuit at busbar 4 from 0.216 to 0.283 s.

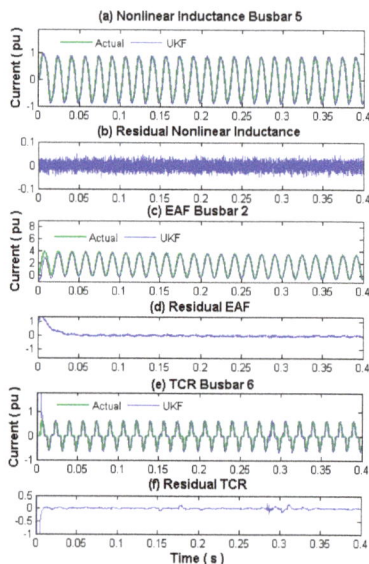

Figure 6. Nonlinear load currents, actual, UKF, and residuals, short-circuit at busbar 4 from 0.216 to 0.283 s.

These state variables show small variations for the considered fault condition. Only in the post-fault period, TCR current differs by approximately 2.5%, but this difference decreases quickly after one cycle to negligible proportions, i.e., approximately to 1%. This error is due to the fast changes in the state variables which make the numerical process of state estimation difficult. The NRMSE between actual and UKF estimated waveforms for nonlinear load currents in Figure 6 gives 0.8% for the nonlinear inductance, 1.35% for the EAF, and 2.16% for the TCR.

4.2. RMS Busbar Voltages under the Short-Circuit Fault at Busbar 4

The voltage sags can be detected directly from the instantaneous or rms values of the nodal voltage waveforms, which are defined as state variables, by comparing the voltage values in the time interval under analysis. If these values vary, a voltage fluctuation (sag or swell) occurs.

Figure 7 shows the rms voltage magnitude for the faulted busbar 4 and for busbars 3, 6, 9, 12 and 14; these busbars are near to busbar 4 and present the largest voltage sags.

The rms magnitude of these voltages is computed using (26), the initial step when the voltage sag begins is due to the short-circuit fault; this time is at cycle 13 or 0.216 s. During the first cycle of post-fault period (cycle 18 or 0.283 s), a noticeable difference is present in the rms voltage of the nearby busbars. The largest difference is 20% for busbar 6, but this error is reduced drastically in the next cycle to 4.5%, being of negligible proportions during the following cycles (approximately 1%). This effect is due to sudden variations in the state variables during and after the fault is removed, which are difficult to follow exactly with the UKF algorithm.

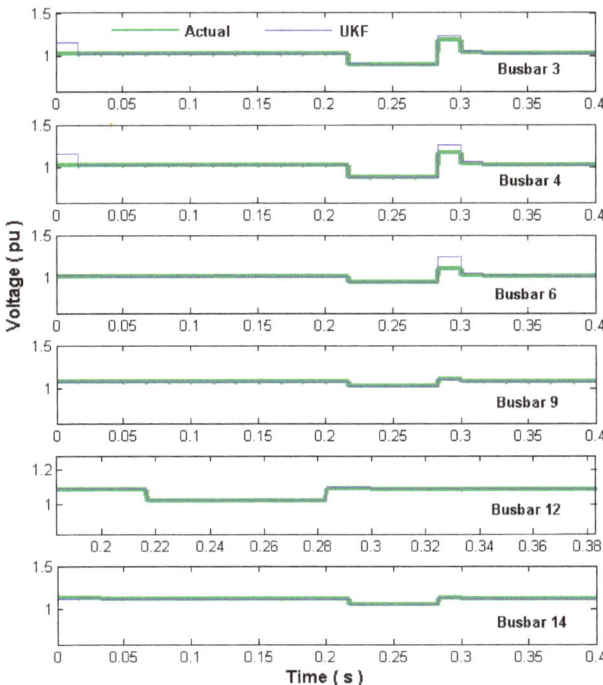

Figure 7. Actual, UKF VSSE, rms voltage magnitude for faulted busbar 4 and busbars 3, 6, 9, 12, and 14. Sags of different magnitude are present from 0.216 to 0.283 s. Swells are present at first post-fault cycle after 0.283 s.

Table 3 shows the actual and estimated voltage sags at the network busbars, referred to the pre-fault magnitudes, due to the single-phase to ground fault at busbar 4. These values are computed again using (26); not listed busbars have a voltage variation of less than 0.01 pu during the fault. The magnitude of the estimated voltage sags closely matches the actual values, thus validating the proposed UKF VSSE methodology.

Table 3. Actual and UKF VSSE voltage sags (pu).

Busbar	Actual	UKF	Busbar	Actual	UKF
3	0.752	0.753	19	0.889	0.890
4	0.713	0.717	20	0.889	0.890
6	0.858	0.860	21	0.892	0.892
7	0.908	0.910	22	0.892	0.893
9	0.870	0.876	23	0.893	0.893
10	0.890	0.900	24	0.887	0.888
12	0.870	0.880	25	0.888	0.889
14	0.880	0.885	26	0.892	0.892
15	0.872	0.873	27	0.891	0.892
16	0.880	0.890	28	0.880	0.881
17	0.892	0.895	29	0.884	0.885
18	0.875	0.880	30	0.907	0.909

4.3. Case Study: UKF VSSE Single-Phase to Ground Fault at Busbar 15

This case study reviews the UKF VSSE when a single-phase to ground fault is applied at busbar 15; the fault impedance is 0.35 pu. This impedance is used to decrease the fault effect in the transient system condition. Busbar 15 has no voltage measurement, however, the state estimation is able to assess its voltage and the voltage of the nearby busbars with the same measurement points of the previous case. Measurements are contaminated with a 2.5% SNR noise. This case study is addressed to represent a short-circuit in the distribution network to assess the VSSE. The state estimation assessment of high power load switching can be also addressed. Figure 8 shows results under the short-circuit fault condition for busbar voltages 15 and 23; these are the busbars that present the largest voltage sag during the examined transient condition.

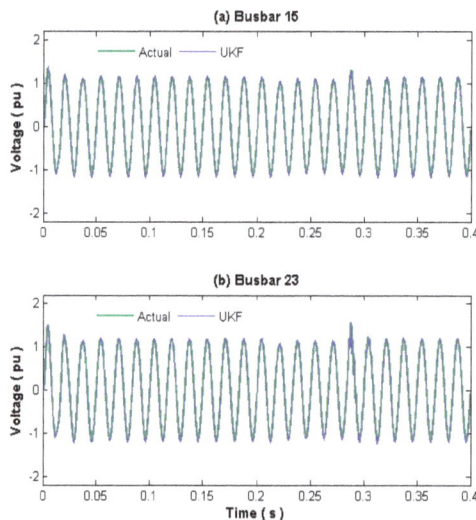

Figure 8. Actual and UKF estimated voltage waveforms of busbars 15 and 23, voltage sag from 0.216 to 0.283 s, cycles 14–17, voltage swell during cycle 18, short-circuit at busbar 15.

A close agreement between the actual and UKF estimated signals including the post-fault period is achieved. Note the swell condition after the fault period. The UKF NRMSE for voltage at busbars 15 and 23 are 1.5%, and 0.65%, respectively.

Figure 9 shows the rms busbar voltages near of the busbar 15. The proposed UKF algorithm gives acceptable estimates for the voltage sag magnitude and duration, mainly for the transient starting and ending time, respectively. This data can be used to classify the type of voltage sags. After the fault period, a voltage swell condition of different magnitude is present during the next two cycles, disappearing when the system transits to its steady state.

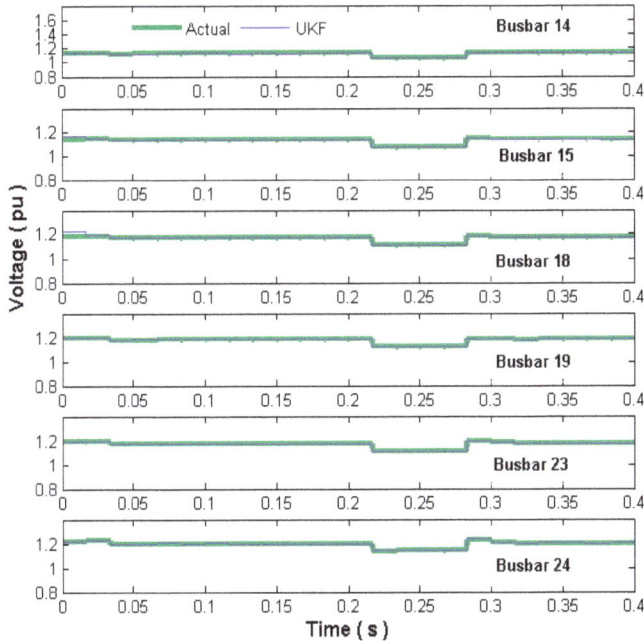

Figure 9. Actual and UKF rms voltage, short-circuit at busbar 15 during 14–17 cycles, from 0.216 to 0.283 s.

4.4. Case Study: UKF VSSE Transient Load Condition at Busbar 24

The proposed UKF-VSSE methodology is applied to estimate a transient load condition; this condition originates a fluctuating voltage sag/swell. The load at busbar 24 varies from cycles 6.25 to 18.75, generating a 12.5 cycle voltage transient in the busbar voltage waveforms. The current demanded by the load at busbar 24 increases 3 times during the first 4.25 cycles of the transient period and 6 times during the next 4 cycles. It then goes back to three times of the initial load current over the following 4.25 cycles, giving a transient condition during 12.5 cycles. Table 4 gives these load changes; the variations may represent mechanical load transients of an electrical motor, the commutation of linear and nonlinear electric loads at the power system busbars, faults, heavy motors starting, or electric heaters turning on, among others. This case study addresses a transient load condition in the distribution system.

Table 4. Transient load condition.

Period	Cycles	Time (s)	Load Current (pu)
Initial	00.00–06.25	0.000–0.104	1.00
Load transient 1	06.25–10.50	0.104–0.175	3.00
Load transient 2	10.50–14.50	0.175–0.241	6.00
Load transient 3	14.50–18.75	0.241–0.312	3.00
Final	18.75–24.00	0.312–0.400	1.00

Figure 10 shows the voltage waveforms at busbars 23, 24, and 25 during the transient load condition. The busbar voltages show the largest fluctuations as a result of the varying load at busbar 24. When the load current increases 3 times, the busbar voltages tend to drop generating a voltage sag. The voltage drops during the first 4.25 cycles of the transient period (6.25 to 10.5 cycles) then again decreases over the next 4 cycles to show the effect of the load current, which increases 6 times during those 4 cycles (10.5 to 14.5 cycles). Finally, the current goes back to three times of the value at the initial period (14.5 to 18.75 cycles).

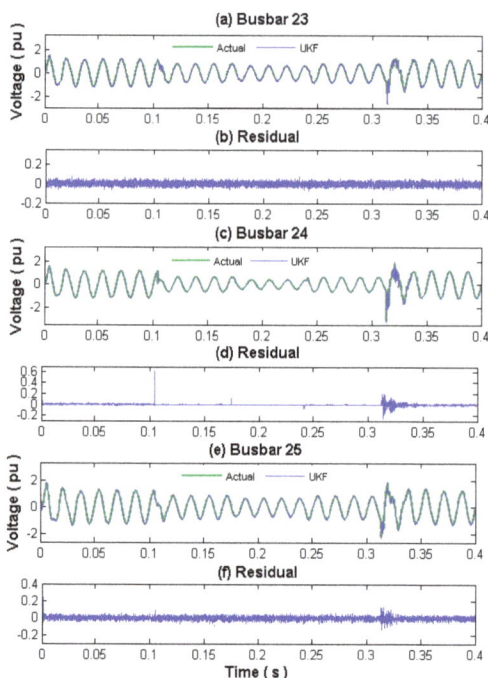

Figure 10. Voltage waveforms, actual, UKF, and residuals of busbars 23, 24, and 25; transient load condition at busbar 24 (0.104 to 0.312 s).

Load transient initiates at 6.25 cycles instead of 6 cycles to evaluate a more critical transient; similarly, the load transient finishes at 18.75 cycles instead of 18 cycles.

The transient state lasts 12.5 cycles (0.208 s), ending at 18.75 cycles (0.312 s), a voltage swell condition is present during the three cycles of the final transient period; voltage waveforms eventually reach the steady state close to the pre-fault operating condition.

NRMSE between actual and UKF estimated waveforms for voltage at busbars 23, 24 and 25 in Figure 10, are 0.45%, 0.40% and 2.43%, respectively.

The rms voltage magnitudes have been calculated using (26) for actual and UKF estimated waveforms during the transient load condition. Figure 11 shows the rms voltage magnitude for each cycle at busbars 21–26, which are close to the load transient of busbar 24.

The obtained rms voltage magnitudes represent the initial, transient and final operation periods, as well as the intermediate transient generating a fluctuating voltage sag. Actual and UKF estimate rms magnitudes closely agree. Please notice the voltage swell of different magnitude during the final period. The proposed UKF VSSE methodology closely estimates these voltage variations.

The use of detailed models to represent the power system components can reduce the state estimation error. Parameters should be close to their real values, filtering the noise from measurements before the assessment of the estimation, and increasing the available measurements.

It should be noted from the above case studies, that the UKF implemented in Matlab script language is still slow to be used in real-time applications. However, with adequate computational techniques such as parallel processing, better computational capability and programs compilation, the execution time can be significantly reduced.

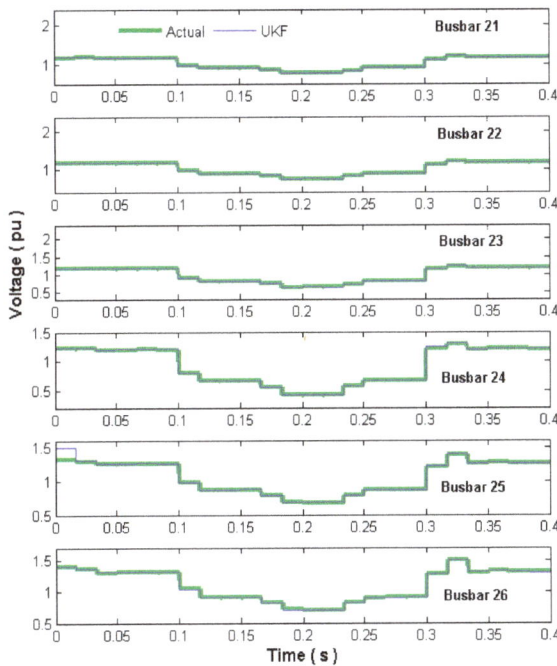

Figure 11. Actual and UKF rms voltage magnitudes, transient load condition at busbar 24 from 0.104 to 0.312 s, during 6.25–18.75 cycles.

5. Conclusions

A time-domain state estimation methodology for voltage sags in power networks using the UKF has been proposed. Nonlinear models for system and measurement equation have been used. It has been demonstrated that the UKF can be applied to precisely assess the voltage sag state estimation in power systems with nonlinear components. The proposed method has been verified using a modified version of the IEEE 30-bus test power system and noisy measurements.

It has been shown that the proposed UKF method dynamically follows the generation of voltage sags, by executing the estimator continuously, to record the voltage sags originated during the power network operation, especially for unmonitored busbars. This requires of an accurate model, a set of synchronized measurements preferably with low noise, sufficient to obtain an observable condition of busbar voltages. The measurement sampling frequency should satisfy the sampling theorem. The rms value can be computed from discrete waveforms; this value gives the information to define the sag magnitude, delimiting the sag time interval.

From the conducted case studies, it has been observed that when the power system goes under fast transients, the UKF estimator error is more noticeable; however, as the network evolves to steady state, the error quickly decreases to negligible proportions, i.e., on average 1%. In most cases, this period is short compared with the voltage sag estimation interval. This condition is present during the final period of the reviewed case studies, when the fault or transient condition is removed. It should be noted that usually at this time, a voltage swell is generated.

The state estimation error increases when sudden transient variations are present. The results obtained with the proposed UKF VSSE methodology have been successfully compared against actual values taken from a simulation of the test power system under the same transient condition. A close agreement has been achieved in all cases between the compared responses.

Author Contributions: R.C.-M. performed the simulation and modelling, analyzed the data, and wrote the paper. A.M. analyzed the results, reviewed the modeling and text, and supervised the related research work. O.A.-L. provided critical comments and revised the paper.

Acknowledgments: The authors gratefully acknowledge the Universidad Michoacana de San Nicolás de Hidalgo through the Facultad de Ingeniería Eléctrica, División de Estudios de Posgrado (FIE-DEP) Morelia, México, for the facilities granted to carry out this investigation. First two authors acknowledge financial assistance from CONACYT to conduct this investigation.

Conflicts of Interest: The authors declare no conflict of interest.

Nomenclature

List of Abbreviations

EAF	Electric arc furnace
FACTS	Flexible alternating current transmission system
KF	Kalman filter
NRMSE	Normalized root mean square error
PQ	Power quality
SNR	Signal to noise ratio
STATCOM	Static synchronous compensator
TCR	Thyristor-controlled rectifier
UKF	Unscented Kalman filter
UT	Unscented transform
VSSE	Voltage sags state estimation
WAMS	Wide area measurement system

List of Symbols

e	State estimation error vector
f	Nonlinear state function
h	Nonlinear output function
k	Time instant $t = k\Delta t$
$k + 1$	Time instant $t = (k + 1)\Delta t$
m	Number of measurements
n	Number of state variables
t	Time vector
u	Input vector
v	Process noise vector
w	Measurement noise vector

x	State vector
$\hat{x}$	Estimated state vector
y	Output vector
z	Measurement vector
E	Expected value
H	Measurements matrix
K	Kalman filter gain matrix
N	Normal distribution
P	Error covariance matrix
Q	Process noise covariance matrix
R	Measurement noise covariance matrix
V_{rms}	Rms voltage magnitude
W	Scalar weights
$+$	A posteriori or after measurement estimate
$-$	A priori or before measurement estimate
Δt	Step time
α	Parameter to determine the spread of sigma points
β	Parameter to determine the distribution of x
λ	Scaling parameter
γ	Scaling parameter
κ	Parameter to determine the spread of sigma points
χ	Sigma points matrix

Appendix A. Per Unit Additional Nonlinear Load Parameters

EAF busbar 2: $L_{eaf} = 0.5$, $k_1 = 0.004$, $k_2 = 0.0005$, $k_3 = 0.005$, $m = 0$, $n = 2.0$, initial condition EAF arc radius = 0.1
Nonlinear inductance busbar 5: $R_m = 4.0$, $L_m = 1.0$, $n = 5.0$, $a = 0$, $b = 0.3$
TCR busbar 6: $R_{tcr} = 1.0$, $L_{tcr} = 0.5$, firing angle $\alpha = 100$ deg.

Appendix B. Nonlinear Models

Appendix B.1. Nonlinear Inductor

Figure A1 shows a nonlinear inductor.

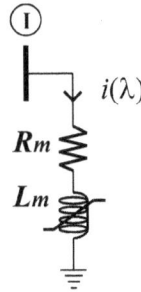

Figure A1. Nonlinear inductance.

According to KVL, the first-order differential equation to represent the nonlinear inductance is:

$$d\lambda/dt = v_I - R_m i(\lambda) \tag{A1}$$

The discrete form of (A1) to define (8) and (9) is given by,

$$\lambda_{(k+1)} = \lambda_{(k)} + \Delta t [d\lambda/dt] \big| k = \lambda_{(k)} + \Delta t \left[v_{I(k)} - R_m i \left(\lambda_{(k)} \right) \right] \tag{A2}$$

where Δt is the time step and k indicates the evaluation at time $t_{(k)}$.

The nonlinear solution of (A1), is represented by $i(\lambda)$, λ is the nonlinear inductor magnetic flux, the polynomial approximation for $i(\lambda)$ is:

$$i(\lambda) = a\lambda + b\lambda^n \tag{A3}$$

n is an odd number due to the odd symmetry of (A3). Coefficients a, b and n adjust the nonlinear saturation curve. The rational fractions and hyperbolic approximations are alternative methods to represent this nonlinearity [38,39].

Appendix B.2. Electric Arc Furnace

Figure A2 shows the EAF model which can be expressed mathematically by two first-order nonlinear differential equations based on the energy conservation law, where the state variables are the arc radius r_{eaf} and the EAF current i_{eaf} [39].

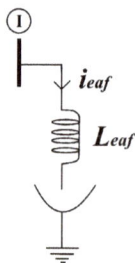

Figure A2. Electric arc furnace.

The first-order nonlinear differential equations to represent the EAF are:

$$dr_{eaf}/dt = (k_3/k_2)r_{eaf}^{(-m-3)}i_{eaf}^2 - (k_1/k_2)r_{eaf}^{(n-1)} \tag{A4}$$

$$di_{eaf}/dt = (1/L_{eaf})\left(v_I - k_3 r_{eaf}^{(-m-2)}i_{eaf}\right), \tag{A5}$$

where n represents the arc cooling effect and m the arc column resistivity [38,39].

The following expressions give the discrete forms of (A4) and (A5) to define (8–9),

$$r_{eaf(k+1)} = r_{eaf(k)} + \Delta t\left[(k_3/k_2)r_{eaf(k)}^{(-m-3)}i_{eaf(k)}^2 - (k_1/k_2)r_{eaf(k)}^{(n-1)}\right], \tag{A6}$$

$$i_{eaf(k+1)} = i_{eaf(k)} + \Delta t\left[(1/L_{eaf})\left(v_{I(k)} - k_3 r_{eaf(k)}^{(-m-2)}i_{eaf(k)}\right)\right] \tag{A7}$$

Appendix B.3. Thyristor Controlled Reactor

A thyristor pair back-to-back connection represents the TCR jointly with an RL circuit. The TCR current is the state variable, the TCR model is shown in Figure A3.

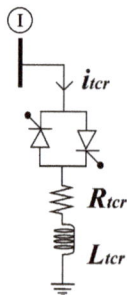

Figure A3. Thyristor controlled reactor.

According to KVL, the first-order nonlinear differential equation modelling the TCR is:

$$di_{tcr}/dt = s(v_I - i_{tcr}R_{tcr})/L_{tcr} \tag{A8}$$

The discrete form of (A8) to define (8) and (9) is given by,

$$i_{tcr(k+1)} = i_{tcr(k)} + \Delta t \left[s_{(k)} \left(v_{I(k)} - i_{tcr(k)}R_{tcr} \right)/L_{tcr} \right] \tag{A9}$$

The TCR current is controlled by the thyristor-firing angle α, the variable s represents this dependency being the switching function to turn on the thyristors, which varies according to the desired firing angle α. This generates harmonic distortion in the voltage and current waveforms. Because of this distortion, the TCR can be considered as a nonlinear component.

References

1. Smith, J.C.; Hensley, G.; Ray, L. *IEEE Recommended Practice for Monitoring Electric Power Quality*; IEEE: Piscataway Township, NJ, USA, 1995; pp. 1159–1995.
2. International Electrotechnical Commission (IEC). *Int. Std. IEC 61000-4-30: Electromagnetic Compatibility (EMC) Part 4-30: Testing and Measurement Techniques—Power Quality Measurement Methods*, 1st ed.; IEC: Geneva, Switzerland, 2003.
3. IEEE Standard. *IEEE Recommended Practice for Evaluating Electric Power System Compatibility with Electronic Process Equipment*; IEEE: Piscataway Township, NJ, USA, 1998; pp. 1346–1998.
4. National Electrical Manufacturers Association. *American National Standard for Electric Power Systems and Equipment-Voltage Ratings (60 Hertz)*; C84.1-2011; National Electrical Manufacturers Association: Washington, DC, USA, 2011.
5. Heydt, G.T. *Electric Power Quality*, 2nd ed.; Stars in a Circle Publications; Lafayette: Paris, France, 1991.
6. Dugan, R.C.; Mcgranaghan, M.F.; Santoso, S.; Wayne, B.H. *Electrical Power Systems Quality*, 2nd ed.; McGraw-Hill: New York, NY, USA, 2002.
7. Sankaran, C. *Power Quality*; CRC Press: Boca Raton, FL, USA, 2002.
8. Bollen, M.H.J. *Understanding Power Quality Problems Voltage Sags and Interruptions*; IEEE Press Series on Power Engineering; IEEE: Piscataway Township, NJ, USA, 2000.
9. Arrillaga, J.; Watson, N.R.; Chen, S. *Power System Quality Assessment*; John Wiley & Sons: Hoboken, NJ, USA, 2000.
10. Watson, N.R. Power quality state estimation. *Eur. Trans. Electr. Power* **2010**, *20*, 19–33. [CrossRef]
11. Yu, K.K.C.; Watson, N.R. An approximate method for transient state estimation. *IEEE Trans. Power Deliv.* **2007**, *22*, 1680–1687. [CrossRef]
12. Medina, A.; Cisneros-Magaña, R. Time-domain harmonic state estimation based on the Kalman filter Poincaré map and extrapolation to the limit cycle. *IET Gener. Transm. Distrib.* **2012**, *6*, 1209–1217. [CrossRef]
13. Cisneros-Magaña, R.; Medina, A. Time domain transient state estimation using singular value decomposition Poincare map and extrapolation to the limit cycle. *Electr. Power Energy Syst.* **2013**, *53*, 810–817. [CrossRef]
14. Espinosa-Juarez, E.; Hernandez, A. A method for voltage sag state estimation in power systems. *IEEE Trans. Power Deliv.* **2007**, *22*, 2517–2526. [CrossRef]
15. Mallick, R.K. Application of linear Kalman filter in power quality estimation. In Proceedings of the ITR International Conference, Bhubaneswar, India, 6 April 2014.
16. Cisneros-Magaña, R.; Medina, A.; Segundo-Ramírez, J. Efficient time domain power quality state estimation using the enhanced numerical differentiation Newton type method. *Electr. Power Energy Syst.* **2014**, *63*, 141–422. [CrossRef]
17. Siavashi, E.M.; Rouhani, A.; Moslemi, R. Detection of voltage sag using unscented Kalman smoother. In Proceedings of the 2010 9th International Conference on Environment and Electrical Engineering (EEEIC), Prague, Czech Republic, 16–19 May 2010; Volume 1, pp. 128–131. [CrossRef]
18. Kusko, A.; Thompson, M.T. *Power Quality in Electrical Systems*; McGraw-Hill: New York, NY, USA, 2007.
19. Fuchs, E.F.; Masoum, M.A.S. *Power Quality in Power Systems and Electrical Machines*; Academic Press Elsevier: New York, NY, USA, 2008.
20. Baggini, A. *Handbook of Power Quality*; John Wiley & Sons: Hoboken, NJ, USA, 2008.

21. Shahriar, M.S.; Habiballah, I.O.; Hussein, H. Optimization of Phasor Measurement Unit (PMU) Placement in Supervisory Control and Data Acquisition (SCADA)-Based Power System for Better State-Estimation Performance. *Energies* **2018**, *11*, 570. [CrossRef]

22. Moreno, V.M.; Pigazo, A. *Kalman Filter: Recent Advances and Applications*; I-Tech Education and Publishing: Vienna, Austria, 2009.

23. Chen, R.; Lin, T.; Bi, R.; Xu, X. Novel Strategy for Accurate Locating of Voltage Sag Sources in Smart Distribution Networks with Inverter-Interfaced Distributed Generators. *Energies* **2017**, *10*, 1885. [CrossRef]

24. Amit, J.; Shivakumar, N.R. Power system tracking and dynamic state estimation. *Power Syst. Conf. Exp.* **2009**. [CrossRef]

25. Wang, S.; Gao, W.; Meliopoulos, A.P.S. An alternative method for power system dynamic state estimation based on unscented transform. *IEEE Trans. Power Syst.* **2012**, *27*, 942–950. [CrossRef]

26. Charalampidis, A.C.; Papavassilopoulos, G.P. Development and numerical investigation of new non-linear Kalman filter variants. *IET Control Theory Appl.* **2011**, *5*, 1155–1166. [CrossRef]

27. Tebianian, H.; Jeyasurya, B. Dynamic state estimation in power systems: Modeling, and challenges. *Electr. Power Syst. Res.* **2015**, *121*, 109–114. [CrossRef]

28. Lalami, A.; Wamkeue, R.; Kamwa, I.; Saad, M.; Beaudoin, J.J. Unscented Kalman filter for non-linear estimation of induction machine parameters. *IET Electr. Power Appl.* **2012**, *6*, 611–620. [CrossRef]

29. Ghahremani, E.; Kamwa, I. Online state estimation of a synchronous generator using unscented Kalman filter from phasor measurements units. *IEEE Trans. Energy Convers.* **2011**, *26*, 1099–1108. [CrossRef]

30. Julier, S.J.; Uhlmann, J.K. Unscented filtering and nonlinear estimation. *Proc. IEEE* **2004**, *92*, 401–422. [CrossRef]

31. Qing, X.; Yang, F.; Wang, X. Extended set-membership filter for power system dynamic state estimation. *Electr. Power Syst. Res.* **2013**, *99*, 56–63. [CrossRef]

32. Huang, M.; Li, W.; Yan, W. Estimating parameters of synchronous generators using square-root unscented Kalman filter. *Electr. Power Syst. Res.* **2010**, *80*, 1137–1144. [CrossRef]

33. Van der Merwe, R.; Wan, E.A. The square-root unscented Kalman filter for state and parameter estimation. In Proceedings of the 2001 IEEE International Conference on Acoustics, Speech, and Signal Processing, Salt Lake City, UT, USA, 7–11 May 2001.

34. Aghamolki, H.G.; Miao, Z.; Fan, L.; Jiang, W.; Manjure, D. Identification of synchronous generator model with frequency control using unscented Kalman filter. *Electr. Power Syst. Res.* **2015**, *126*, 45–55. [CrossRef]

35. Bretas, N.; Bretas, A.; Piereti, S. Innovation concept for measurement gross error detection and identification in power system state estimation. *IET Gener. Transm. Distrib.* **2011**, *5*, 603–608. [CrossRef]

36. Jain, S.K.; Singh, S.N. Harmonics estimation in emerging power system: Key issues and challenges. *Electr. Power Syst. Res.* **2011**, *81*, 1754–1766. [CrossRef]

37. University of Washington, Electrical Engineering, Power Systems Test Case Archive. Available online: http://www.ee.washington.edu/research/pstca/pf30/pg_tca30bus.htm (accessed on 15 March 2018).

38. Waston, N.R.; Bathurst, G. Task Force on Harmonics Modeling and Simulation, Modeling devices with nonlinear voltage-current characteristics for harmonic studies. *IEEE Trans. Power Deliv.* **2004**, *19*, 1802–1811. [CrossRef]

39. Acha, E.; Madrigal, M. *Power Systems Harmonics Computer Modelling and Analysis*; John Wiley & Sons: Hoboken, NJ, USA, 2001.

energies

MDPI

Article

Research on Modeling of Microgrid Based on Data Testing and Parameter Identification

Junjun Zhang [1,2,*], Yaojie Sun [1,*], Meiyin Liu [2], Wei Dong [2] and Pingping Han [3,*]

[1]   College of Information Science and Technology, Fudan University, Shanghai 200433, China
[2]   State Key Laboratory of Operation and Control of Renewable Energy & Storage, China EPRI,
        Nanjing 210003, China; liumeiyin@epri.sgcc.com.cn (M.L.); dongwei@epri.sgcc.com.cn (W.D.)
[3]   College of Electrical Engineering and Automation, Hefei University of Technology, Hefei 230009, China
*   Correspondence: zhangjunjun@epri.sgcc.com.cn (J.Z.); yjsun@fudan.edu.cn (Y.S.);
        LH021211@163.com (P.H.); Tel.: +21-55665508 (Y.S.); +551-62901417 (P.H.)

Received: 31 July 2018; Accepted: 19 September 2018; Published: 21 September 2018

Abstract: The model parameter identification based on real operation data is a means to accurately determine the simulation parameters of the microgrid, but the real operation data cannot guarantee the exact agreement with the required data for parameter identification, which has become an important restriction factor in the accurate simulation and analysis of the dynamics of the microgrid. This paper provides a method of modeling of microgrid based on data testing and parameter identification. In this paper, the method of parameter trajectory sensitivity is first introduced. Then, the data testing scheme for parameter identification is presented, and the parameter identification flow chart is given. Thirdly, a microgrid demonstration system in China is taken as an example, the important parameters of the distributed photovoltaic, direct-drive wind turbine and energy storage unit in the system are obtained by data testing and parameter identification, and in the end, the accuracy of the model is verified through the comparison of the simulation data and the test data of the microgrid during grid-connection/island switching process. The obtained microgrid model provides a base model for the analysis of the overall characteristics, such as the transient stability, as well as power quality of the microgrid.

Keywords: modeling method; parameter identification; data testing; microgrid; grid-connection/ island switching process

---

1. Introduction

A microgrid with multiple distributed renewable power sources is conducive to increasing the application of renewable energy power generation, reducing energy consumption, and improving the reliability and flexibility of power systems, thus, is developing rapidly worldwide. At present, there are more than 400 microgrid demonstration projects planned, under construction or put into operation worldwide.

Digital modeling and simulation is one of the main methods of microgrid research, which provides a necessary tool and strong technical support for the study of microgrid operation mechanisms, protection control and other issues. Models developed by manufacturers of wind turbines, photovoltaic (PV) units, and energy storage units can reproduce the behavior of distributed power sources accurately and in detail. However, this level of detail is not suitable for the stability study of large power systems. This is because the use of these models requires a lot of input data, and due to the large number of state variables in the models, the time and complexity of simulations are greatly increased. Therefore, the standards committees in China and abroad have issued general dynamic models for renewable power sources such as wind turbines, PV, energy storage, etc. On one hand, the general models can accurately simulate the dynamic behavior of renewable power supply, on the other hand, it is suitable for large-scale power grid research [1–4].

With the general models that provide a unified modeling standard for power system analysis, the accuracy of model parameters will directly affect the results of the system analysis, because when the power grid is disturbed, it may cause large economic losses. At present, the following problems exist in obtaining model parameters: Firstly, the manufacturer is unwilling to provide accurate parameters of the model for reasons of confidentiality; secondly, the general model used in system analysis is not completely consistent with the model provided by the manufacturer; thirdly, the model parameters presented by the manufacturer are applied to ideal working conditions of a single component. When multiple components are running at the same time, the interaction between the components, and the interference of various factors during real operation, cause the output result of the model to have errors with the result under the original ideal working conditions. Therefore, determining how to obtain the key parameters in the simulation model is crucial to the accuracy of the system analysis results.

Model parameter identification based on real operation data is currently a method that can accurately determine the simulation parameters. The parameter identification methods of power system mainly include: The time domain identification method, the frequency domain identification method and the intelligent optimization method. The time domain identification method directly identifies model parameters based on the time domain sampling information of the system, and the most commonly used method is the least square method [5]. The frequency domain identification method uses the fast Fourier transform to convert the time domain information of the system into the frequency domain, and obtains the parameter value through the frequency domain response function of the system [6]. The intelligent optimization method is based on the global optimization characteristics of the optimization algorithm itself, and determines the optimal value of the model parameters by calculating the fitness of the objective function. Intelligent optimization algorithms that are popularly used include the ant colony algorithm [7], particle swarm algorithm [8] and genetic algorithm [9].

Parameter identification with simulation data for the study of algorithms is practical, such as references [10–14]. Although, only the parameters identified from the real operation data can be used in the simulation analysis of the actual system [15–19]. However, the real operation data obtained online cannot guarantee the exact agreement with the required data for parameter identification. Therefore, based on the above research, the data testing scheme of an actual microgrid for parameter identification has been presented in this paper, and the data testing scheme and parameter identification are applied to a microgrid demonstration system in Jiangsu province, China for a more accurate simulation model. Characteristics of the grid-connection/island switching process of the microgrid are tested and simulated to verify the effectiveness of the model.

The structure of the paper is as follows: The second part introduces the parameters to be identified, the data testing scheme, the flow chart of parameter identification and the validation index based on parameter trajectory sensitivity analysis. The third part gives the identification results of the actual microgrid and verifies the accuracy of the model by comparing the simulation data and the test data during the grid-connection/island switching process. Finally, the paper gives the conclusions and the issues to be studied in the future.

2. Parameter Identification and Data Testing for Microgrid Based on Sensitivity Analysis

2.1. Parameter Trajectory Sensitivity

The trajectory sensitivity of parameters is a measure of the difficulty of parameter identification. The greater the trajectory sensitivity, the greater the influence of parameters on the dynamic behavior of the system and the easier it is to identify [20]. Trajectory sensitivity (relative value) is defined as the following formula:

$$S_{\theta_i} = \lim_{\Delta\theta_i \to 0} \frac{\frac{y(t,\theta_1,...,\theta_i+\Delta\theta_i,...,\theta_m)-y(t,\theta_1,...,\theta_i,...,\theta_m)}{y(t,\theta_1,...,\theta_i,...,\theta_m)}}{\frac{\Delta\theta_i}{\theta_{i0}}} \tag{1}$$

where S_{θ_i} is the trajectory sensitivity of the i-th parameter θ_i, and θ_{i0} is the given value of the parameter θ_i; y is one of the observed quantities. m is the number of parameters to be identified.

Research shows that if the observed quantity is sensitive to the parameters, that is, the corresponding trajectory sensitivity is relatively large, the parameters can be easily identified according to the observed quantities. On the contrary, if the parameter has little influence on all the observed quantities, then the parameter is not easy to identify. In order to quantitatively compare the sensitivity of each parameter, the average value of the absolute values of each point of the track sensitivity can be calculated according to the following formula:

$$A_{S_{\theta_i}} = \frac{1}{K} \sum_{K=1}^{K} |S_{\theta_i}| \tag{2}$$

where k is the total number of points of trajectory sensitivity.

2.2. Determination of Parameters to be Identified and Data Testing Scheme for the Microgrid

2.2.1. Sensitivity Analysis

For the PV, direct-drive wind turbine and energy storage unit in the microgrid discussed in this paper, the converter is the core component. So it is a key technology for distributed generation unit modeling to establish a reasonable model of the converter and its controller and to obtain its parameters accurately.

In this paper, based on the general models discussed in Section 1, and according to the analysis of transient characteristics of microgrids, models of the PV, direct-drive wind turbine and energy storage unit control system are established, respectively.

First, for the PV unit, the output of the photovoltaic cell is simulated by an equivalent DC source, and PV has both an MPPT operating mode and a constant power mode. The control structure of the PV unit is shown in Figure 1. The parameters that need identification include the maximum and minimum of power change rate when the PV unit works in MPPT mode and PI (Proportion Integration) parameters of active and reactive power control links. This is because in MPPT mode, a sudden change in the active power output may damage the converter, and a slow change may be harmful to the recovery of the system. The PI control parameters of active and reactive power control links directly affect the dynamic characteristics of PV unit. The parameters to be identified are shown in the red dashed box in Figure 1.

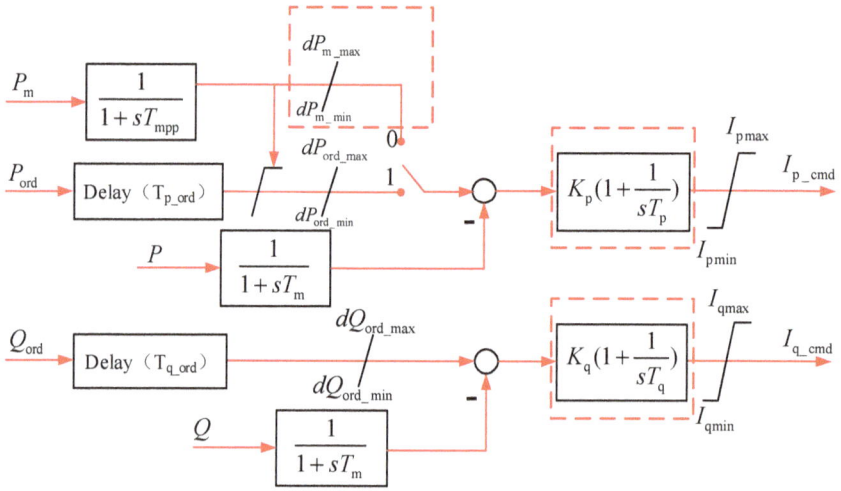

Figure 1. Control structure diagram of the photovoltaic (PV) unit.

For the trajectory sensitivity analysis, a grid-connected PV system model is built on MATLAB/Simulink platform. The active power P and reactive power Q at the outlet of the PV unit are selected as the observation measurements, sensitivity analysis is done to K_p, T_p, K_q and T_q, all of which are parameters of the PI controllers of the PV unit in Figure 1 and need identification.

In Figure 1, $K_p(1 + \frac{1}{sT_p})$ is the general form of PI controller, where, K_p represents the ration coefficient, T_p represents the integral time constants. However, in the simulation model constructed in MATLAB, the form of PI control is $K_p(1 + \frac{K_{ip}}{s})$, where K_{ip} is used to describe the integral link, and the relationship between K_{ip} and T_p is: $K_{ip} = K_p/T_p$. Similarly, the relationship between K_{iq} and T_q is: $K_{iq} = K_q/T_q$. Due to the MATLAB platform, the trajectory sensitivity analysis given in Figures 3 and 5 adopts K_{ip} and K_{iq} instead of T_p and T_q.

(1) Trajectory sensitivity when power reference value changes.

The reference values of the active power and reactive power of the converter are changed according to Figure 2a,b respectively. The trajectory sensitivity of the parameters to be identified is shown in Figure 3a,b.

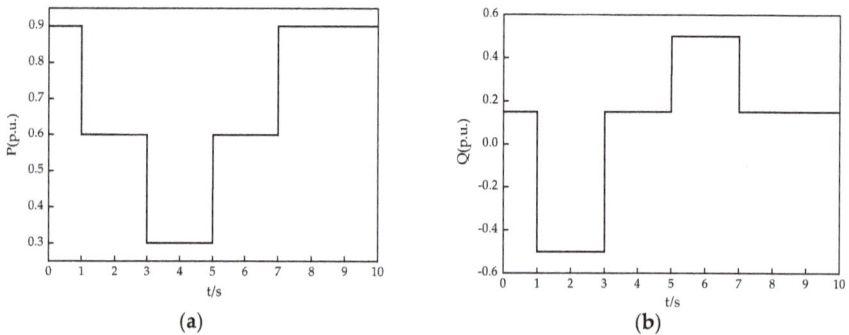

(a)

(b)

Figure 2. (a) Reference value of active power of the converter; (b) reference value of reactive power of the converter.

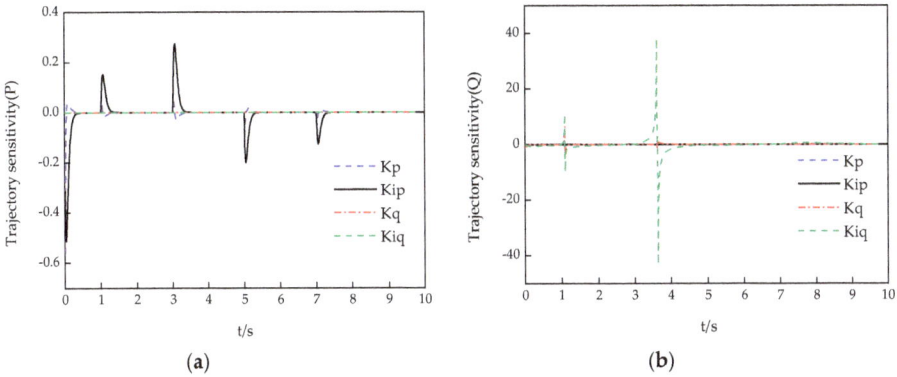

Figure 3. (a) Sensitivity curve of active power of the parameter to be identified; (b) sensitivity curve of reactive power of the parameter to be identified.

(2) Trajectory sensitivity in the case of short-circuit fault.

At the point of common coupling (PCC), a three-phase short-circuit fault occurs at t = 0 s, and the voltage drops to 0.8 p.u. At t = 0.15 s, the fault is cleared, and the system gradually resumes normal operation. The voltage (U) at PCC is shown in Figure 4. The trajectory sensitivity of the parameters to be identified is shown in Figure 5a,b.

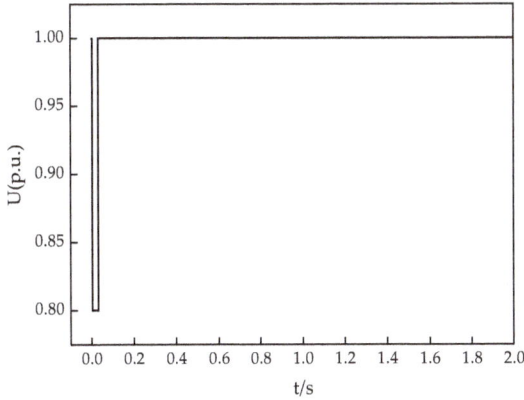

Figure 4. Reference value of the voltage at the point of common coupling (PCC).

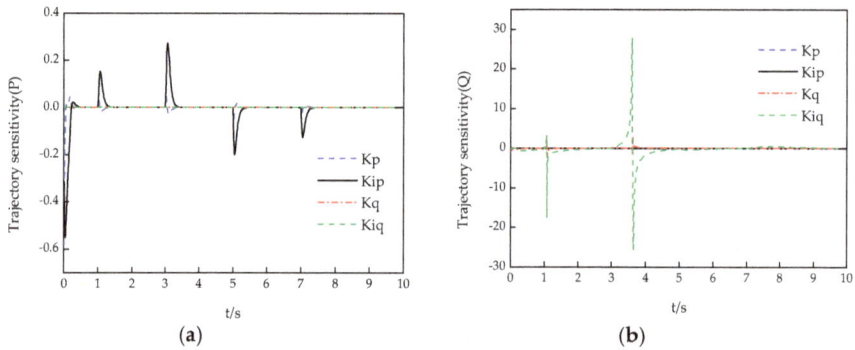

Figure 5. (**a**) Sensitivity curve of active power of the parameter to be identified; (**b**) sensitivity curve of reactive power of the parameter to be identified.

The sensitivity analysis results show that, under the conditions of both reference value change and the short-circuit fault, active power is more sensitive to the parameters K_p and T_p, and reactive power is more sensitive to the parameters K_q and T_q. Therefore, with an appropriate data testing scheme, the dynamic data of the output power of the distributed generation unit under different working conditions can be collected to identify the required parameters.

(3) Sensitivity of upper and lower limits of power change rate.

The acquisition of the upper and lower limits of the power change rate does not require sensitivity analysis, but only needs to set the power change reference value in the actual test process and calculate the power change rate according to the test data.

(4) Sensitivity analysis of direct-drive wind turbine.

For direct-drive wind turbine, the machine-side variables and the grid-side variables are decoupled by the DC link. Only the PI control parameters of DC voltage and reactive power control link of the grid side converter directly affect its dynamic characteristics. Therefore, the machine-side part is replaced by an equivalent DC source, and the control structure is shown in Figure 6. The parameters that need identification include PI parameters of the DC voltage control link and the reactive power control link. The parameters to be identified are shown in the red dashed box in Figure 6.

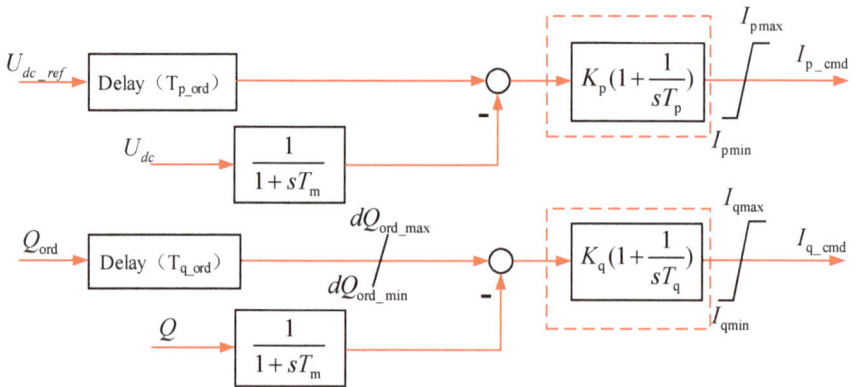

Figure 6. Control structure diagram of direct-drive wind turbine.

Adopting the sensitivity analysis method used in the PV unit, the sensitivity analysis is undertaken for the direct-drive wind turbine. The results show that the DC voltage of the direct-drive wind turbine is more sensitive to the parameters K_p and T_p, and the reactive power is more sensitive to the parameters K_q and T_q.

(5) Sensitivity analysis of energy storage unit.

For the energy storage unit, the output of the energy storage battery is also simulated by an equivalent DC source, of which the control structure is shown in Figure 7. In the grid-connected mode of a microgrid, the energy storage unit works in power control mode, and normally the Q command is 0. Therefore, the parameters that need identification include the upper and lower limits of the active power control slope and the PI control parameters of the active and reactive power control links. The parameters to be identified are shown in the red dashed box in Figure 7.

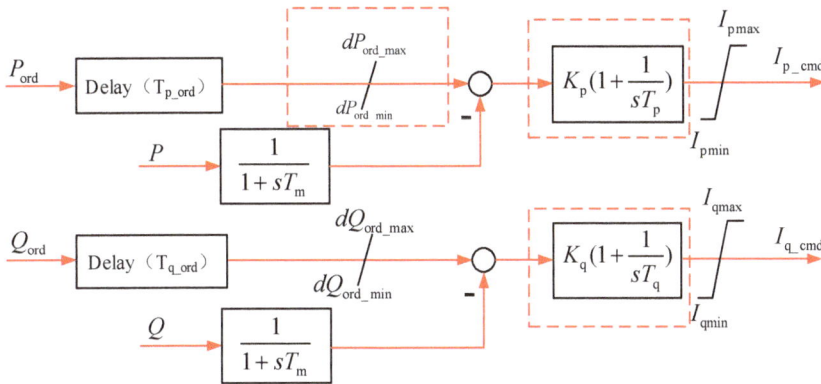

Figure 7. Control structure diagram of energy storage unit.

Adopting the sensitivity analysis method used in the PV unit, the sensitivity analysis is undertaken for the energy storage unit. The result show that the active power of the energy storage unit is more sensitive to the parameters K_p and T_p, and the reactive power is more sensitive to the parameters K_q and T_q.

2.2.2. Test Scheme

The sensitivity analysis gives the correlation between system parameters and system external characteristics and helps to determine the observation measurements for parameter identification. According to the sensitivity analysis in the previous section, the test scheme was designed, as shown in Table 1.

Table 1. Parameters to be identified and test scheme of the renewable power sources.

Test Object	Parameters to Be Identified	Test Scheme	Operation Mode
PV unit	Parameters of maximum power tracking control link dP_{m_max}, dP_{m_min}	Test of the DC source input disturbance	MPPT
	Parameters of power control link K_p, T_p, K_q, T_q	Test of voltage disturbance	Power control
Direct-drive wind turbine	Parameters of power control link K_p, T_p, K_q, T_q	Test of voltage disturbance	Power control
Energy storage unit	Parameters of power slope control link $dP_{ord\_max}, dP_{ord\_min}$	Test of power command value disturbance	Power control
	Parameters of power control link K_p, T_p, K_q, T_q	Test of voltage disturbance	Power control

2.3. Parameter Identification Steps Based on Particle Swarm Optimization Algorithm

Considering the nonlinearity of the system to be identified, the time cost and the accuracy of the algorithm [21–23], the particle swarm optimization algorithm is adopted for parameter identification. The flow chart is shown in Figure 8.

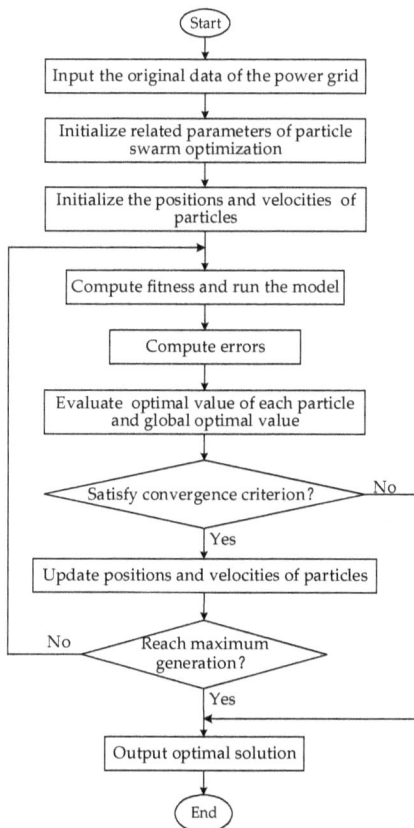

Figure 8. Flow chart of parameter identification.

2.4. Error Indicator

According to the China Recommended Standard GB/T 32892-2016 "photovoltaic power generation system model and parameter testing procedures", the maximum deviation between the simulation data and the test data during the steady-state interval is adopted to assess the accuracy of the model. The expression is as follows:

$$F = \max_{i=K_{M_{Start}}...K_{M_{End}}} (|X_S(i) - X_M(i)|) \tag{3}$$

where X_S is the standard value of the electrical quantity in the simulation data; X_M is the standard value of the electrical quantity in the test data; $K_{M_{Start}}$ and $K_{M_{End}}$ are the first and last serial numbers of the test data in the error calculation interval.

3. Example Analysis

3.1. Basic Information and Selection of Test Points

The modeling and test work are carried out on the wind/PV/storage microgrid demonstration system established by the Electric Power Research Institute of State Grid Jiangsu Electric power corporation. The system is a microgrid infrastructure platform which includes a distributed wind/PV/storage unit, i.e., one 30 kW horizontal axis direct-drive wind turbine, one 5 kW rotor wing direct-drive wind turbine, one 5 kW H airfoil direct-drive wind turbine, one 30 kW fixed polycrystalline PV unit and one 75 kWh lithium battery energy storage unit. The detailed parameters of the microgrid example system are shown in Appendix A.

The wind turbines, PV power unit and energy storage unit of the microgrid are connected in parallel to a 0.4 kV bus, which connects to a 10 kV distribution network for power exchange with the main grid.

The test points are selected according to the sensitivity and test scheme determined in Table 1. The test basis is as follows: GB/T 32826-2016 "photovoltaic power generation system modeling guide", GB/T 32892-2016 "photovoltaic power generation system model and parameter test regulation", and GB/T 34133-2017 "energy storage converter test technical regulation".

The test points of the microgrid in the example are shown in Figure 9. The test equipment connection points are shown in red. A total of four test points are used to test the voltage and current of the PV unit, wind turbines, energy storage unit as well as the grid-connection terminal respectively.

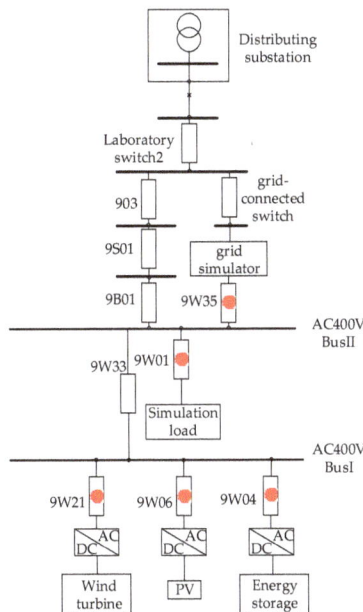

Figure 9. Electrical wiring diagram of modeling simulation and test of the microgrid.

3.2. Effectiveness Analysis of the Microgrid Model

3.2.1. Parameter Identification Result

Using the method described in Section 2, the parameters of the microgrid demonstration system are identified, and the results are shown in Table 2.

Table 2. Parameter identification results of the wind/PV/energy storage unit in the microgrid.

Name	Definition	Identification Result of PV Unit	Identification Result of 30 kW Horizontal Axis Wind Turbine	Identification Result of 5 kW Rotor Wing Wind Turbine	Identification Result of 5 kW H Wing Wind Turbine	Identification Result of Energy Storage Unit
dPm_max	Upper limit of maximum power tracking slope (pu/s)	151.64	Not included in the test			
dPm_min	Lower limit of maximum power tracking slope (pu/s)	52.1				
Kp	PI control proportional coefficient of the active power control	0.6372	1.55	1.72	0.77	0.53
Tp	PI control integral time constant of the active power control (s)	0.0049	0.0083	0.0013	0.0013	0.0063
Kq	PI control proportional coefficient of the reactive power control	1.1962	1.33	1.34	0.077	0.68
Tq	PI control integral time constant of the reactive power control (s)	0.0157	0.0022	0.0013	0.0054	0.0026
dPord_max	Upper limit of active power control slope (pu/s)	Not included in the test				12.46
dPord_min	Lower limit of active power control slope (pu/s)					3.87

3.2.2. Validity Analysis of the Microgrid Model

The microgrid has two modes: Grid-connected mode and island mode. In the normal operation of the main power grid, the microgrid works in the grid-connected mode. If a fault occurs in the main power grid, the relay protection device of the power system acts, and the connection between the microgrid and the main power grid is disconnected. The microgrid enters the island operating state. During the microgrid grid-connection/island switching process, frequency and voltage fluctuations of the distribution network and the microgrid may occur.

In order to verify the accuracy of the overall model of the microgrid, the grid-connection/island switching tests and simulation analysis were carried out. The test data and the simulation data were compared to verify the validity of the parameter identification results. The test cases and test points are shown in Table 3.

Table 3. Grid-connection/island switching test of the microgrid.

Test Cases	Test Points
Control from grid-connected mode to island mode	AC side of PV and energy storage unit; PCC
Control from island mode to grid-connected mode	AC side of PV and energy storage; PCC

PCC refers to point of common coupling.

Case 1: Normally, the microgrid works in grid-connected mode, the power supplied by the PV unit is 20 kW, energy storage unit supplies 50 kW. Then, at t = 10 s, the microgrid is switched to island mode from grid-connected mode. The comparison of the simulation results and the test data are shown in Figure 10. The errors are shown in Table 4, the fulfillment of the results is obtained according to GB/T 32892-2016 "photovoltaic power generation system model and parameter test regulation". The data includes: The voltage, active power and reactive power at PCC, the active power supplied by the energy storage unit and the PV unit respectively.

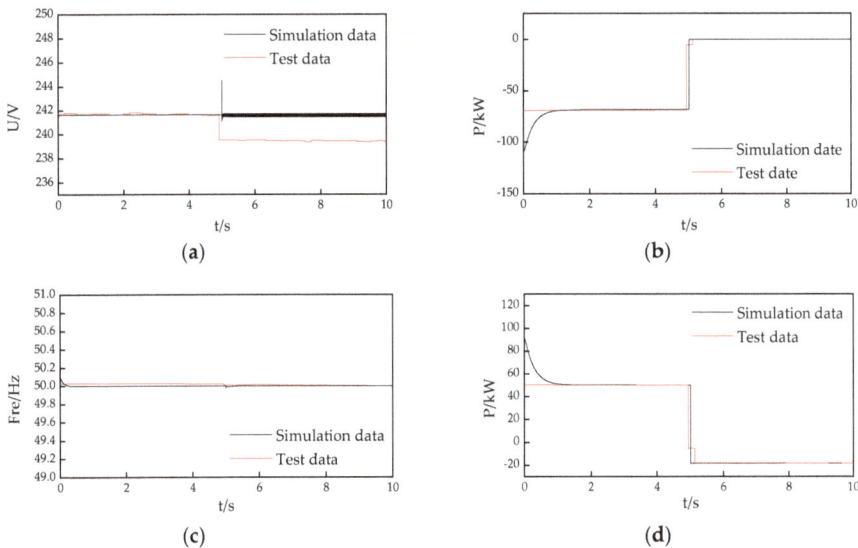

Figure 10. (a) Comparison of the voltage at PCC; (b) Comparison of the active power at PCC; (c) Comparison of the frequency at PCC; (d) Comparison of the active power of energy storage unit.

Table 4. Errors in case 1 condition.

Name	F	Fulfill the Regulation
Voltage at PCC	0.024	Yes
Frequency at PCC	0.0006	Yes
Active power at PCC	0.029	Yes
Active power of the energy storage unit	0.016	Yes
Active power of PV unit	0.024	Yes

Case 2: Normally, the microgrid works in island mode, the frequency and voltage are controlled by the energy storage unit. The frequency is 50 Hz. The power supplied by the PV unit is 20 kW, and the load is 32 kW. Then, at t = 2.2 s, the microgrid is switched to grid-connected mode from island mode. The comparison of the simulation results and the test data are shown in Figure 11. The errors are shown in Table 5.

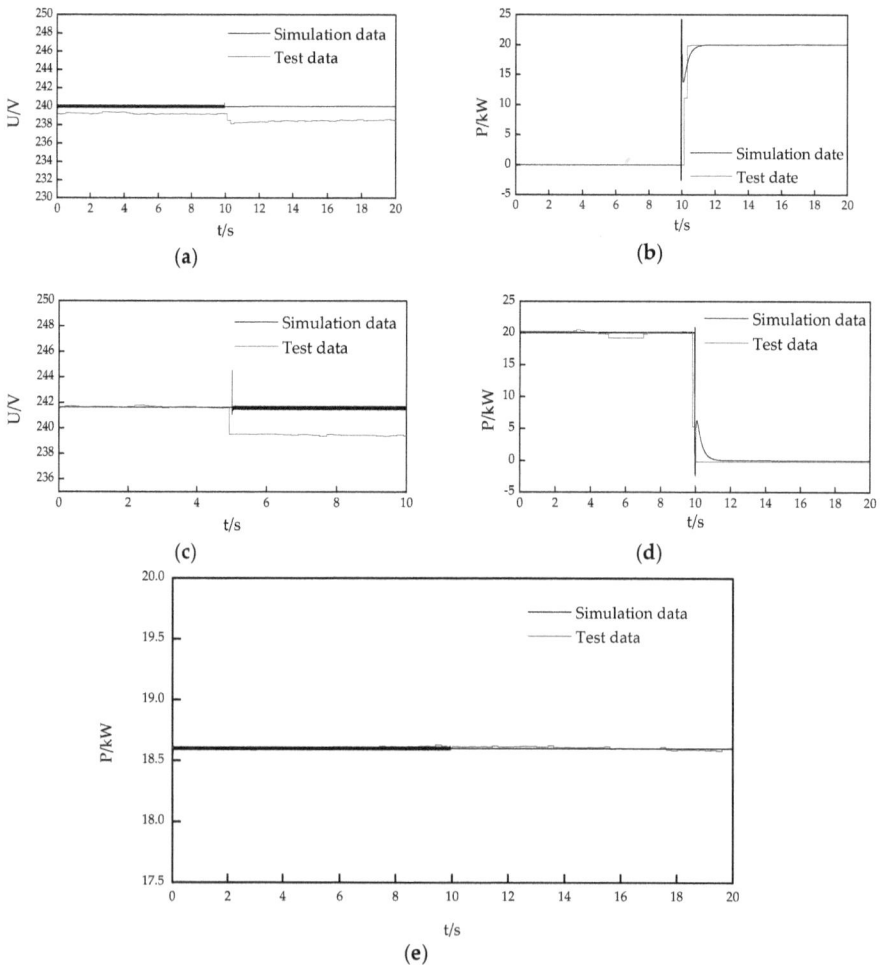

Figure 11. (a) Comparison of the voltage at PCC; (b) Comparison of the active power at PCC; (c) Comparison of the frequency at PCC; (d) Comparison of the active power of energy storage unit; (e) Comparison of the active power of PV unit.

Table 5. Errors in case 2 condition.

Name	F	Fulfill the Regulation
Voltage at PCC	0.023	Yes
Frequency at PCC	0.00012	Yes
Active power at PCC	0.030	Yes
Active power of energy storage unit	0.011	Yes
Active power of PV unit	0.0047	Yes

3.3. Summary

According to the simulation results, the frequency and voltage are within normal limits during the transient process of grid-connection/island switching. In addition, the simulation model meets the grid guide in terms of error indicators, which provides a reasonable model for the analysis of the overall operating characteristics of the microgrid, and thus help guarantee the stable operation of the microgrid and its practical application in China.

4. Conclusions

In view of the influence of manufacturers, models, and other factors on the accuracy of parameters in microgrid modeling, this paper proposes an accurate modeling method for microgrids, which provides a model basis for the analysis of the overall characteristics of a microgrid and may help popularize the application of microgrids. The conclusions obtained in this paper and issues that require further study include:

(1) The key to accurate modeling of microgrids is that the actual data used for parameter identification are consistent with the actual system operating conditions. Based on the results of the parameter sensitivity analysis, the paper designed a data test scheme to ensure the validity of the data used in the parameter identification.

(2) Taking the microgrid demonstration system in China as an example, the important parameters of the PV, wind turbine, and energy storage unit in the system were obtained, and the accuracy of the model was verified through the comparison of the simulation data and the test data during the grid-connection/island switching process. Therefore, it can be concluded that the modeling method proposed in this paper is effective.

(3) The general model of renewable power generation units were adopted in the paper. When studying the special control structure of the microgrid, the corresponding test scheme and algorithm need to be further improved.

(4) In the microgrid control system, the current parameter identification focuses on the parameters in power and frequency control. Future studies should pay more attention to the increasing demand of low voltage ride through controls of the microgrid.

(5) The workload of parameter identification based on test data is large. When the number of microgrids increases, the workload increases exponentially. It is necessary to develop specialized and commercial software to improve the efficiency of microgrid parameter identification and modeling.

Author Contributions: This paper was a collaborative effort between the authors. The authors contributed collectively to the collation and review of literatures.

Acknowledgments: This paper is based on the China National Key Research and Development Plan "Smart Grid and Equipment" (2016YFB0900600) and Project Supported by Science and Technology Foundation of State Grid Corporation of China (52110417000F).

Conflicts of Interest: The authors declare no conflict of interest.

Appendix A

Table A1. Basic information of the microgrid of Electric Power Research Institute of State Grid Jiangsu Electric power corporation.

Microgrid Composition			Technical Parameter	
Wind Turbine	Horizontal Axis Wind Turbine 30 kW	Manufacturer	Shanghai Ghrepower Green Energy Co., Ltd.	
		Type/Quantity	GNW36k3G/1 set	
		Rated Power	36 kW	
		Rated Voltage	380 V ± 15%	
	Rotor-Type Wind Turbine 5 kW	Manufacturer	Shanghai Linfeng Wind Energy Technology Co., Ltd.	
		Type/Quantity	1 set	
		Rated Power	5 kW	
		Rated Voltage	380 V	
	H-Type Vertical Axis Wind Turbine 5 kW	Manufacturer	Shanghai Linfeng Wind Energy Science and Technology Co., Ltd.	
		Type/Quantity	P5000-AB/1 set	
		Rated Power	5 kW	
		Rated Voltage	380 V	
Photovoltaic Inverter		Manufacturer	Sungrow Power Supply Co., Ltd.	
		Type/Quantity	SG30K3/1 set	
		Rated Power	30 kW	
		Rated Voltage	380 V	
Energy Storage Converter		Manufacturer	Jiangsu Fangcheng Electric Science and Technology Co., Ltd.	
		Type/Quantity	1 set	
		Rated Power	500 kW	
		Rated Voltage	380 V	
Information about the connection to the power network		The voltage level of access point	10 kV	
		Power supply distance of the grid	/	
		The location of point of common coupling	Distributed microgrid base platform 10 kV power distribution room	
		Short circuit capacity	Positive sequence short circuit capacity of bus three-phase short circuit in maximum operation mode 289 MVA	
		Whether there is a step-up transformer or not	☑Yes	☐ No
			Quantity: 1	Capacity per unit: 500 kVA

References

1. Han, P.P.; Lin, Z.H.; Wang, L.; Fan, G.J.; Zhang, X.A. A Survey on Equivalence Modeling for Large-Scale Photovoltaic Power Plants. *Energies* **2018**, *11*, 1463. [CrossRef]
2. Xu, S. Modeling and Simulation of Hybrid AC and DC Power System Considering Multiple Distributed Energy Sources. *Modern Electr. Power* **2017**, *35*, 32–38.
3. Liu, C.; Shi, W. Comparison of China's Wind Power Integration Standard with Similar Foreign Standards. *Smart Grid* **2014**, *2*, 48–51.
4. Chi, Y.N.; Zhang, Z.K.; Li, Y.; Wei, L.J. Development of Large-scale Wind Power Grid Integration and Technical Standard. *North China Electric Power* **2017**, 1–7. [CrossRef]
5. Xiong, X.F.; Chen, K.; Zheng, W.Z.; Shen, Z.J.; Shahzad, N.M. Photovoltaic Inverter Model Identification Based on Least Squares Method. *Power Syst. Prot. Control* **2012**, *40*, 52–57.
6. Liu, L.F. Parameter Identifieation of Generators for Large Power System Dynamic Equivalence. Master's Thesis, Wuhan University, Wuhan, China, 2004.
7. Dong, W.; Li, Y.; Xiang, J. Optimal Sizing of a Stand-Alone Hybrid Power System Based on Battery/Hydrogen with an Improved Ant Colony Optimization. *Energies* **2016**, *9*, 785. [CrossRef]
8. Malik, S.; Kim, D.H. Prediction-Learning Algorithm for Efficient Energy Consumption in Smart Buildings Based on Particle Regeneration and Velocity Boost in Particle Swarm Optimization Neural Networks. *Energies* **2018**, *11*, 1289. [CrossRef]

9. Zhu, M.X.; Li, J.C.; Chang, D.G.; Zhang, G.J.; Chen, J.M. Optimization of Antenna Array Deployment for Partial Discharge Localization in Substations by Hybrid Particle Swarm Optimization and Genetic Algorithm Method. *Energies* **2018**, *11*, 1813. [CrossRef]
10. Liu, C.L.; Liang, W.P.; Sun, W.Y.; Su, J. Model Identification of Thermal Process in Power Plant. In Proceedings of the 2004 IEEE Region 10 Conference TENCON 2004, Chiang Mai, Thailand, 24 November 2004.
11. Shi, L.J.; Zhao, C.Y.; Wu, F.; Zhao, H.Y.; Shi, J.J.; Jiang, X.X. Optimization of ESS PI Controllers Parameters Based on PSO in Smart Grid. In Proceedings of the 2015 5th International Conference on Electric Utility Deregulation and Restructuring and Power Technologies (DRPT), Changsha, China, 26–29 November 2015; pp. 2566–2570.
12. Shen, W.J.; Li, H.X. A Sensitivity-Based Group-Wise Parameter Identification Algorithm for the Electric Model of Li-ion Battery. *IEEE Access* **2017**, *5*, 4377–4387. [CrossRef]
13. Yang, W.; Shan, L.; Qu, Y.; Yang, B.; Sun, J. Co-Simulation and Parameters Identification Algorithm Research of Servo system based on dynamic module. In Proceedings of the 2015 International Conference on Control, Automation and Information Sciences (ICCAIS), Changshu, China, 29–31 October 2015; pp. 408–413.
14. Li, Z.; Zhen, S.; Zhen, X.Q.; Bao, W. System Identification of Adaptive Reduced Order Based on PSO Algorithm. *Ind. Control Comput.* **2017**, *30*, 112–115.
15. Xu, Y.; Gao, Z.; Zhu, X.R. Research on Parameter Identification of Photovoltaic Array Based on Measured Data. In Proceedings of the 2017 20th International Conference on Electrical Machines and Systems (ICEMS), Sydney, NSW, Australia, 11–14 August 2017; pp. 1–5.
16. Yang, H.C.; Cheng, R.F.; Lv, C.Y.; Wang, X.W. Study on Internal Parameter Identification and Output Characteristics of Photovoltaic Module. *Appl. Electron. Tech.* **2018**, *44*, 125–128. [CrossRef]
17. Xu, Y.; Gao, Z.; Zhu, X.R. Multi-scenario Parameters Identification of Photovoltaic Array Based on Hybrid Artificial Fish Swarm and Frog Leaping Algorithm. *Renew. Energy* **2018**, *36*, 519–526. [CrossRef]
18. Chen, C. Research on the Identification of Synchronous Generator Parameters Based on Measured System Disturbance. Master's Thesis, South China University of Technology, Guangzhou, China, 2017.
19. Li, P.Q.; Li, H.; Li, X.R. Optimized Identification Strategy for Composite Load Model Parameters Based on Sensitivity and Correlation Analysis. *Trans. China Electrotech. Soc.* **2016**, *31*, 181–188. [CrossRef]
20. Pan, X.P.; Yin, Z.H.; Jv, P.; Wu, F.; Jin, Y.Q.; Ma, Q. Model Parameter Identification of DFIG Based on Short Circuit Current. *Electr. Power Autom. Equip.* **2017**, *37*, 27–31. [CrossRef]
21. Dong, Z.; Han, P.; Wang, D.F.; Jiao, S.M. Thermal Process System Identification Using Particle Swarm Optimization. *IEEE Int. Symp. Ind. Electron.* **2006**, *1*, 194–198. [CrossRef]
22. Xing, H.; Pan, X.J. Application of Improved Particle Swarm Optimization in System Identification. In Proceedings of the 2018 Chinese Control and Decision Conference (CCDC), Shenyang, China, 9–11 June 2018; pp. 1341–1346.
23. Sonia, S.; Hari, M.P. Genetic Algorithm, Particle Swarm Optimization and Harmony Search: A quick comparison. In Proceedings of the 2016 6th International Conference-Cloud System and Big Data Engineering (Confluence), Noida, India, 14–15 January 2016; pp. 40–44.

energies

MDPI

Article

Control-Loop-Based Impedance Enhancement of Grid-Tied Inverters for Harmonic Suppression: Principle and Implementation

Fei Wang *, Lijun Zhang, Hui Guo and Xiayun Feng

College of Mechatronics Engineering and Automation, Shanghai University, Shanghai 200444, China; lijunzhang@shu.edu.cn (L.Z.); guohui0827@163.com (H.G.); arthur@shu.edu.cn (X.F.)
* Correspondence: f.wang@shu.edu.cn; Tel.: +86-21-6613-6638

Received: 10 September 2018; Accepted: 18 October 2018; Published: 23 October 2018

Abstract: To understand different control loops that have been proposed to improve the quality of current into grid from the perspective of output impedance, control-loop-based output impedance enhancement of grid-tied inverters for harmonic suppression is proposed in this paper. The principle and generalized control loop deduction are presented for reshaping the output impedance. Taking a traditional LCL (Inductor-Capacitor-Inductor)-type inverter with dual-loop control as an example, different kinds of control loop topologies are derived step by step and further optimized for the implementation of the proposed principle. Consequently, the improved control consists of a filtering-capacitor voltage loop, and a grid current loop is found which can remove the existing inner capacitor current loop and therefore simplify the control. Finally, the effectiveness of the proposed control method is compared with the existing method and both are verified by simulations and experiments.

Keywords: grid-tied inverter; harmonic suppression; impedance enhancement; output impedance

1. Introduction

This paper considers the suppression of current harmonics from the perspective of output impedance reshaping. In this section, the background of harmonic suppression, literature review, formulation of the problem of interest for this investigation, scope and contribution of this study, and organization of the paper are provided.

1.1. Background and Significance

With the wide application of power-electronics-based distributed generator equipment, the quality of current into grid is partly polluted with the harmonics caused by the process of switching and the distorted grid voltage [1]. The loss on the cables will increase and some sensitive loads will work abnormally or even break if large harmonic contents of current exist, leading to the low economy of a power system. Therefore, to suppress the current harmonics as well as to achieve the goal of grid-tied requirements for inverters, different kinds of schemes for harmonic suppression have been proposed, including different regulators, topologies, and control loops [2].

1.2. Literature Review

Originally, the L-type grid-tied inverter is usually employed in a distributed generation system with a single grid-current feedback control loop. To satisfy the grid-tied requirements on the harmonics current, a large inductor is generally adopted to mainly suppress the harmonics at the switching frequency. However, the scale of the inverters is large and the effect of harmonic suppression is not sufficient. Therefore, to reduce the scale of inverters as well as achieve a better effect on suppressing current harmonics, an LCL filter is employed [3]. However, as the LCL filter has a resonance peak, the inverter will become unstable if the grid voltage or the inverter has an excitation source at the resonant frequency. To damp the resonant peak, the dual-loop control strategy with capacitor current and grid current has been proposed [4]. The positive resonance peak is compensated by a negative resonance peak which is introduced though an additional compensator [5,6]. However, since the capacitor current contains a large amount of high-frequency harmonics, it is difficult to compensate them accurately. Therefore, instead of using a capacitor current, the weighted average current flowing through the two inductors of an LCL filter is used. This method is a split formation of capacitor current because the capacitor current is equal to the difference of an inverter-side current and grid-side current for an LCL filter [7,8].

Besides the harmonics caused by switches, the background harmonics of grid voltage also affect the grid current [9,10]. To compensate the harmonics in the grid voltage, the grid voltage is directly feedforward (FF), i.e., an equivalent voltage source which has the same magnitude and phase with grid voltage is generated to attenuate the impact of grid voltage disturbance. This method is called the partial feedforward control strategy. It is more suitable for the L-type structure, although for the LCL-type structure, due to the role of the capacitor, partial feedforward control cannot fully compensate the background harmonics of grid voltage [11,12]. The magnitude and phase of the generated equivalent voltage will not be equal to the grid voltage through an LCL filter if partial feedforward control strategy is used. Therefore, the coefficient of feedforward should be modified to get full compensation, that is, a full feedforward control strategy [13]. Other control loops, such as in [14,15], which employ the combination of current feedback and grid voltage feedforward also achieve the effect of suppressing the current harmonics. In addition, some advanced control algorithms such as repetitive control and the adjoint method can be taken into consideration for harmonic suppression [16,17].

1.3. Formulation of the Problem of Interest for This Study

It is necessary that the guideline of harmonic suppression from the perspective of reshaping output impedance should be summarized since the different kinds of control loops which have been proposed in previous literatures are intrinsically enhancing the output impedance. Generally, the equivalent model of grid-tied inverter systems is equal to a current source in parallel with an output impedance [18]. If the output impedance is large enough within the control bandwidth, the current source will become as ideal as possible. Therefore, based on the idea of infinite output impedance enhancement, this paper investigates the principle regarding reshaping the output impedance and optimizing the control loops.

1.4. Scope and Contribution of This Paper

In this paper, through summarizing the prior different control loops, the concept of output impedance enhancement is put forward, thus providing an understanding of different control loops. An optimized control loop is derived from the grid voltage feedforward control strategy, which can remove the existing inner capacitor current loop and therefore simplify the control loop. Meanwhile, the capability of output impedance enhancement is identical to the grid voltage feedforward control strategy. Besides, the cost of the system is reduced and its practical implementation is more convenient.

1.5. Organization of the Mauscript

The rest of this paper is organized as follows: by reshaping the output impedance based on the concept of output impedance enhancement, the grid voltage feedback and grid current feedback methods of reshaping output impedance are obtained in Section 2. The impact of two basic methods on output impedance enhancement is analyzed in detail in Section 3. To simplify the control loop for the convenience of realization, a method of splitting is adopted to optimize the grid voltage feedforward control loop. Then, a novel control loop is proposed in Section 4. The correctness of the proposed control loop is verified in simulation and experiment in Section 5. Finally, a conclusion is made in Section 6.

2. Principle of Control-Loop-Based Impedance Enhancement

2.1. Reshaping Principle Based on Output Impedance Models

The LCL-type grid-tied inverter with dual loop of capacitor current and grid current is employed. The structure is shown in Figure 1a, where U_{dc}, u_{inv}, u_C, u_{PCC}, and u_g, represent DC voltage, voltage at inverter side, capacitor voltage, voltage at point of common coupling, and grid voltage, respectively. i_{inv}, i_C, i_g, i_{ref}, I_g, and i_{Cref} represent current at inverter side, capacitor current, grid current, reference current, the magnitude of the reference current, and the reference capacitor current, respectively. L_1, L_2, C_f, and Z_g represent the filter inductor at the inverter side, filter inductor at grid side, filter capacitor, and grid impedance, respectively. G_{ig} and G_{ic} represent the grid current regulator and the capacitor current regulator, respectively.

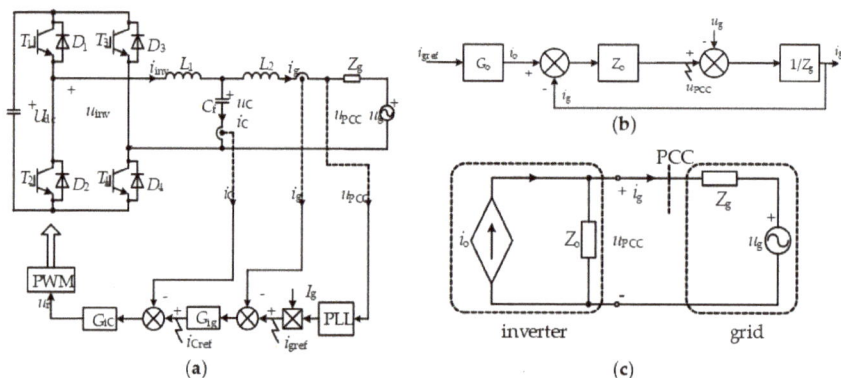

Figure 1. LCL (Inductor-Capacitor-Inductor)-type grid-tied inverter for single phase: (**a**) structure of main circuit; (**b**) structure of impedance model; (**c**) structure of equivalent circuit model.

i_{ref} and i_o are satisfied with $i_o = i_{gref}G_o$, where G_o is a transfer function of current gain. Figure 1c shows the equivalent circuit model. It consists of a controlled current source in parallel with an output impedance and a voltage source in series with a grid impedance [6].

According to Figure 1c, i_g can be expressed as

$$i_g = \frac{Z_o}{Z_o + Z_g}i_o - \frac{1}{Z_o + Z_g}u_g. \tag{1}$$

It is found in (1) that i_g is affected by i_o, u_g, Z_o, and Z_g. If Z_o is increased, the second item of (1) will be decreased. Therefore, the impact of u_g on i_g will be reduced by enhancing Z_o. If Z_o is increased to be much larger than Z_g, that is, $Z_o \gg Z_g$, (1) will be simplified to

$$i_g = i_o. \tag{2}$$

It can be found that the impact of grid voltage disturbance on grid current can be reduced by increasing the output impedance, thereby improving the quality of current into grid. Therefore, i_g is approximately affected by i_o but not affected by u_g and Z_g.

The methods of reshaping output impedance are listed in Figure 2. Take Figure 2a as an example: an extra impedance Z_{os} is in series with grid impedance. The output impedance will be enhanced if Z_{os} is designed properly. Other reshaping ways are shown in Figure 2b,h.

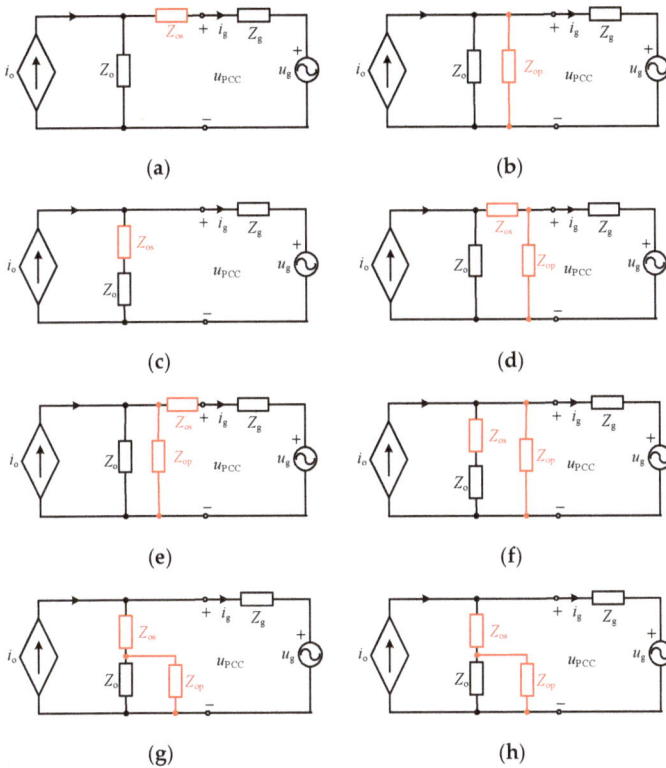

Figure 2. Methods of reshaping output impedance based on equivalent circuit: (**a,c**) series reshaping; (**b**) parallel reshaping; (**d–h**) combination of series and parallel reshaping.

2.2. Deduction Principle Based on Output Impedance Models

If an extra impedance Z_{os} is added to the grid side (shown in Figure 2a), the control structure will be changed into Figure 3a. Compared to Figure 1b, a grid current feedback loop is added to the control loop. To conveniently make a further control, the reference current i_{ref} is chosen, since a single point that has physical meaning is needed. Then, Figure 3a can be transferred to Figure 3b. It is obvious that the two structures are an equivalent formation. Similarly, the other reshaping ways shown in Figure 2b–h can be transferred to Figure 4b–h, respectively.

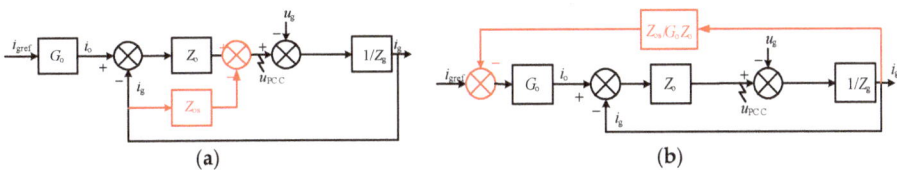

Figure 3. Block diagram of reshaping output impedance: (**a**) impedance model; (**b**) deformed impedance model.

Figure 4b,c,f,g, are grid voltage feedback control loops which feed grid voltage back to the reference current point. Figure 4a is a grid current feedback control loop which feeds the grid current back to the reference current, and Figure 4d,e,h are the combinations of grid voltage and grid current feedback control loops.

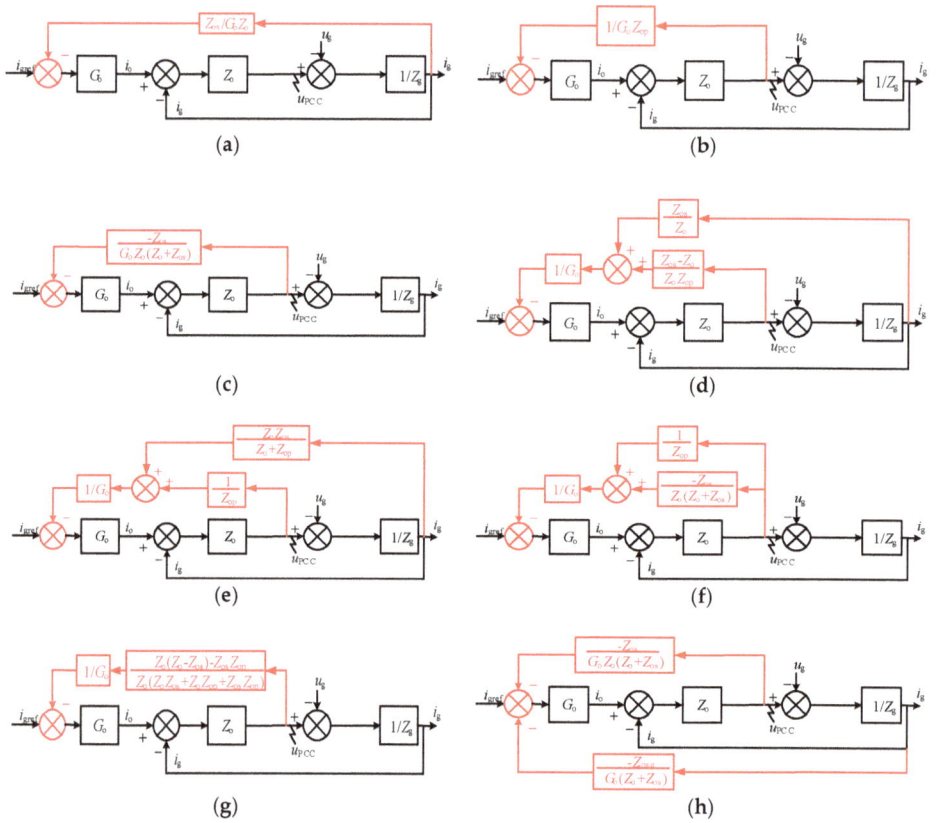

Figure 4. Methods of reshaping output impedance based on control loops: (**a,c**) series reshaping; (**b**) parallel reshaping; (**d–h**) combination of series and parallel reshaping.

From Figure 4a–h, two basic methods are obtained which can be employed to reshape output impedance: (a) grid voltage feedback control strategy and (b) grid current feedback control strategy. In the next section, the impact of output impedance based on two basic methods will be analyzed.

3. Analysis of Impact on Output Impedance of Various Control

3.1. Impact on Output Impedance Using Voltage Feedback Control

For convenience, K_u is adopted to uniformly represent the feedback coefficient when the grid voltage feedback control is employed. K_u can be expressed either as a constant value or a transfer function. Its control diagram is shown in Figure 5.

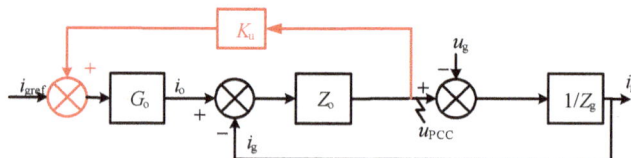

Figure 5. The structure of grid voltage feedback control.

The new output impedance Z_{ou} is

$$Z_{ou} = \frac{u_{PCC}}{(-i_g)}\bigg|_{i_{gref}=0} = Z_o - \frac{Z_o}{K_uG_oZ_o - 1}K_uG_oZ_o = \frac{Z_o}{1 - K_uG_oZ_o}. \tag{3}$$

Considering G_0 is positive in general within the control bandwidth, Z_{ou} can be enhanced when K_u is positive.

3.2. Impact on Output Impedance Using Current Feedback Control

K_i represents the feedback coefficient when the grid current feedback control is employed. The control diagram is shown in Figure 6.

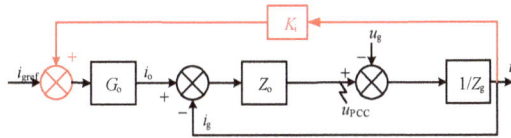

Figure 6. The structure of grid current feedback control.

The new output impedance Z_{oi} is

$$Z_{oi} = \frac{u_{PCC}}{(-i_g)}\bigg|_{i_{gref}=0} = Z_o - K_iG_oZ_o. \tag{4}$$

As indicated in (4), Z_{oi} will be enhanced when K_i is negative. If (5) is satisfied, the two basic methods will have an identical ability to enhance output impedance. Bode plots are given in Figure 7 to verify the enhancement effectiveness of the output impedance.

$$Z_{ou} = Z_{oi} \rightarrow K_i = \frac{K_uZ_o}{K_uG_oZ_o - 1}. \tag{5}$$

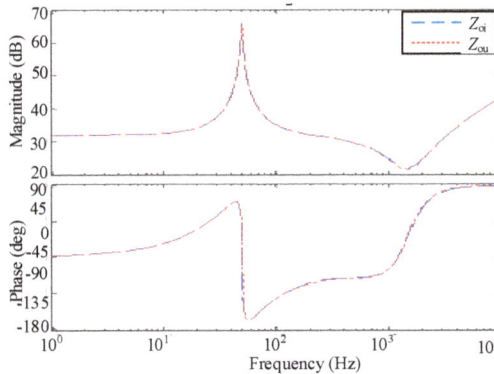

Figure 7. Bode plots of output impedance with two basic impedance reshaped methods.

(a). As indicated in (4), Z_{oi} will be enhanced when K_i is negative. If (5) is satisfied, the two basic methods will have an identical ability to enhance output impedance. Bode plots are given in Figure 7 to verify the enhancement effectiveness of the output impedance. The output impedance will be enhanced when K_u is positive for voltage feedback control, while K_i is negative for current feedback control.

(b). The capacity of the two basic methods to reshape the output impedance is identical and they are equivalent when (5) is satisfied except in one circumstance, that is, when K_u is equal to $1/G_oZ_o$, K_i will trend to be infinite.

To obtain the largest gain of output impedance in order to prove that the basic impedance reshaped methods have the instructive significance for the enhancement of output impedance, a specific example is illustrated.

3.3. Impedance Enhancement Control Loop Based on Basic Impedance Reshaped Methods

Take the dual loop with capacitor current and grid current feedback as an example. When a grid voltage feedback control loop is added to the reference current, as shown in Figure 8a, the control diagram can be obtained (Figure 8b).

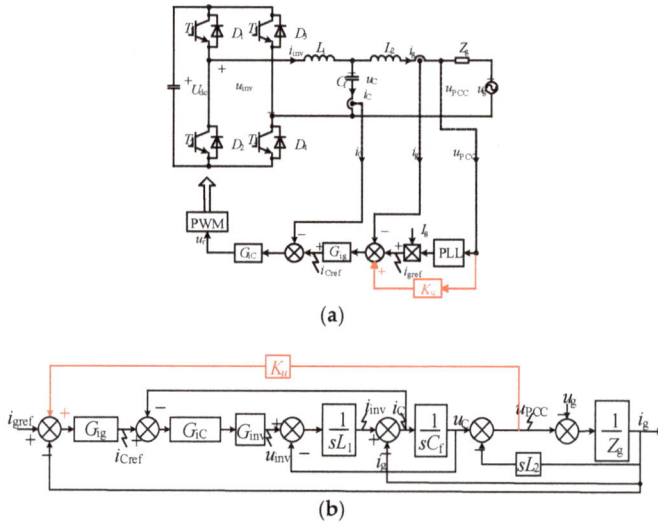

(a)

(b)

Figure 8. LCL-type grid-tied inverter for single-phase: (**a**) the main circuit structure with grid voltage feedback control loop; (**b**) control loop.

According to Figure 8b, the original output impedance without grid voltage feedback control loop is

$$Z_o = \frac{s^3 L_1 L_2 C_f + s^2 G_{iC} G_{inv} L_2 C_f + sL_1 + sL_2 + G_{ig} G_{iC} G_{inv}}{s^2 L_1 C_f + s G_{iC} G_{inv} C_f + 1}. \tag{6}$$

G_o can be calculated as

$$G_o = \frac{i_g}{i_{ref}}\Big|_{u_{PCC}=0} = \frac{G_{ig} G_{iC} G_{inv}}{s^3 L_1 L_2 C_f + s^2 G_{iC} G_{inv} L_2 C_f + sL_1 + sL_2 + G_{ig} G_{iC} G_{inv}}. \tag{7}$$

The reshaped output impedance can be calculated according to

$$Z_{ou} = \frac{Z_o}{1 - K_u G_o Z_o} = \frac{s^3 L_1 L_2 C_f + s^2 G_{iC} G_{inv} L_2 C_f + sL_1 + sL_2 + G_{ig} G_{iC} G_{inv}}{s^2 L_1 C_f + s G_{iC} G_{inv} C_f + 1 - K_u G_{ig} G_{iC} G_{inv}}. \tag{8}$$

If the denominator of (8) is zero, the output impedance can be enhanced as much as possible. Then, K_u is

$$K_u = \frac{s^2 L_1 C_f + s G_{iC} G_{inv} C_f + 1}{G_{ig} G_{iC} G_{inv}}. \tag{9}$$

Simplify (9) and the deformed feedback coefficient K'_u is

$$K'_u = \frac{s^2 L_1 C_f + s G_{iC} G_{inv} C_f + 1}{G_{inv}}. \tag{10}$$

In order to reduce high-frequency noise introduced by the second-order differential item K'_u, a first-order low pass filter is employed. The output impedance will turn into

$$Z'_{ou} = \frac{Z_o}{1 - \frac{K'_u}{T_{LPF}s+1} \times G_o Z_o} = \frac{s^3 L_1 L_2 C_f + s^2 G_{iC} G_{inv} L_2 C_f + s L_1 + s L_2 + G_{ig} G_{iC} G_{inv}}{s^2 L_1 C_f + s G_{iC} G_{inv} C_f + 1 - \frac{K'_u}{T_{LPF}s+1} \times G_{ig} G_{iC} G_{inv}} \tag{11}$$

where T_{LPF} is the time constant of the low-pass filter. The voltage harmonics are mainly considered about less than 40th, and the cut-off frequency of the low-pass filter should be larger than 2 kHz in order to enhance output impedance as much as possible. So, $T_{LPF} = 40$ μs. (Cut-off frequency is 3980 Hz, approximately).

The parameters of an LCL inverter system are given in Table 1. According to (8), (11), and Table 2, the original output impedance and reshaped output impedance can be obtained, as shown in Figure 9. Figure 9 shows that the reshaped output impedance is obviously enhanced under the cut-off frequency, and the highest gain is about 1000 times greater than before.

Table 1. Parameters of single phase inverters with LCL filter.

Symbol	Quantity	Value
u_g	Grid voltage	220 V
f_g	Frequency	50 Hz
U_{dc}	DC voltage	400 V
L_1	Filter inductor at inverter side	2.4 mH
L_2	Filter inductor at grid side	2.4 mH
Z_g	Grid impedance	0.2 Ω/0.08 mH
C_f	Filter capacitor	4 μF
f_s	Switch frequency	16 kHz
G_{iC}	Capacitor current regulator	30
G_{inv}	Gain of the inverter	1
K_p	Proportional coefficient	1
K_R	Resonance regulator coefficient	50
ω_c	Band frequency	10 rad s^{-1}
T_{LPF}	Time constant of low pass filter	40 μs
i_{ref}	Reference current	10 A

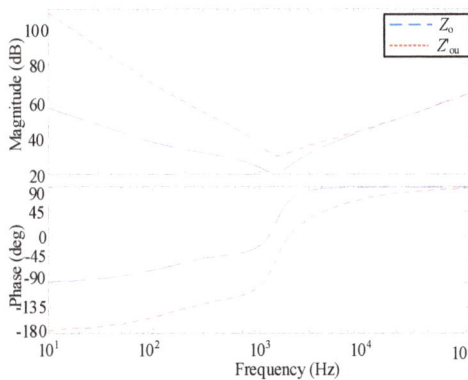

Figure 9. Bode plot of output impedance based on voltage feedback control.

Table 2. THD of grid current under different harmonic suppression schemes.

Control Strategy	THD of Grid Current/%
Dual-loop control strategy	13.17
Grid voltage feedforward control strategy	5.78
Proposed control strategy	5.75

Actually, this method is also called the full feedforward control strategy [9]. The principle of reshaped output impedance is elaborately described in previous sections. Based on the impedance model, two basic methods of reshaping output impedance are obtained, and a specific example is illustrated to prove the effectiveness of the basic methods. Further, it is necessary to apply the methods to engineering implementation.

4. Implementation and Optimization of Control-Loop-Based Impedance Enhancement

According to the Kirchhoff voltage law, grid voltage is equal to the difference between capacitor voltage and grid-side inductor voltage, and the expression of the feedback branch can be transformed to

$$K'_{\mathrm{u}} = \frac{s^2 L_1 C_{\mathrm{f}} + s G_{\mathrm{iC}} G_{\mathrm{inv}} C_{\mathrm{f}} + 1}{G_{\mathrm{inv}}}. \tag{12}$$

The expression of the feedback branch turns from $u_{\mathrm{PCC}} K_{\mathrm{u}}$ to $(u_C - u_{L2}) K_{\mathrm{u}}$, which means the control loop turns from a single loop of grid voltage to a dual loop of capacitor voltage and inductor voltage. If the feedback coefficient is equal to K_{u}, the capacity of reshaping output impedance is identical to the grid voltage feedforward control. The control structure is shown in Figure 10. K_Z is the feedback coefficient of the grid-side inductor voltage. A new control loop structure is created through a split. Similarly, the inductor voltage at grid side is also equal to the grid current multiplied by the inductor impedance, and then the grid voltage also can be transformed to

$$u_{\mathrm{PCC}} K_{\mathrm{u}} = (u_C - i_{\mathrm{g}} Z_{L_2}) K_{\mathrm{u}} \tag{13}$$

where Z_{L2} represents inductor impedance. The control structure is shown in Figure 11. K_{I} is the feedback coefficient of the grid current.

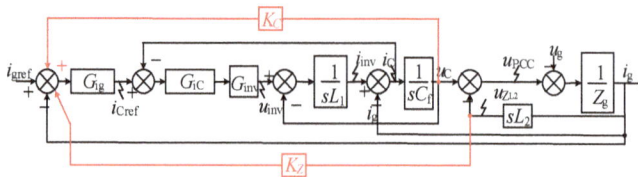

Figure 10. The structure of capacitor voltage and grid-side inductor voltage feedback control loop.

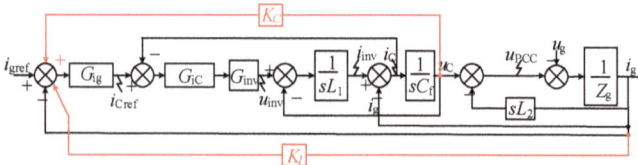

Figure 11. The structure of capacitor voltage and grid current feedback control loop.

To obtain a simpler control loop, continue to simplify to

$$
\begin{aligned}
u_{PCC}K_u &= (u_C - i_g Z_{L2})K_u = u_C K_u - i_g Z_{L2}K_u = u_C\left(\frac{s^2 L_1 C_f + s G_{iC}G_{inv}C_f + 1}{G_{ig}G_{iC}G_{inv}}\right) - i_g s L_2\left(\frac{s^2 L_1 C_f + s G_{iC}G_{inv}C_f + 1}{G_{ig}G_{iC}G_{inv}}\right) \\
&= u_C\frac{s^2 L_1 C_f + 1}{G_{ig}G_{iC}G_{inv}} + u_C\frac{s C_f}{G_{ig}} - i_g\left(\frac{s^3 L_1 L_2 C_f + s^2 G_{iC}G_{inv}C_f L_2 + s L_2}{G_{ig}G_{iC}G_{inv}}\right) \\
&= u_C\frac{s^2 L_1 C_f + 1}{G_{ig}G_{iC}G_{inv}} + i_C\frac{1}{G_{ig}} - i_g\left(\frac{s^3 L_1 L_2 C_f + s^2 G_{iC}G_{inv}C_f L_2 + s L_2}{G_{ig}G_{iC}G_{inv}}\right)
\end{aligned}
\tag{14}
$$

Compared to the FF control strategy, the capacitor voltage, capacitor current, and grid current feedback control loops are employed instead of the grid voltage feedback control loop as shown in Figure 12a. The control structure in Figure 12a can be optimized to Figure 12b through simplifying the control loops where the new controller G'_{ig} equals $G_{ig}G_{iC}$. G'_{ig} is a proportional-resonant controller similar to G_{ig} in the FF control strategy. Otherwise, the proposed control method is derived from the FF control strategy, and the capability of enhancing output impedance regarding the two methods is identical, as is the control bandwidth. The specific control structure of the proposed method is shown in Figure 12c. It shows that the inner capacitor current loop is eliminated and the capacitor current is no longer needed compared to the FF control strategy. Therefore, the cost will be reduced and the implementation in practice will be convenient.

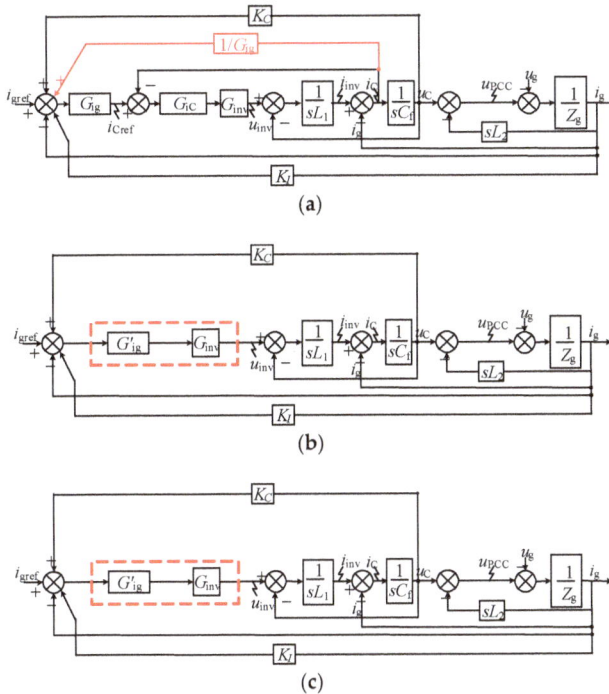

Figure 12. Proposed control loop of capacitor voltage and grid current feedback. (**a**) Optimization step I; (**b**) optimization step II; (**c**) proposed control structure, respectively.

To deal with harmonic influences of the grid voltage on the grid current, the capacitor feedforward control loop is proposed from the perspective of enhancing output impedance. The output impedance is enhanced in the full bandwidth compared to the multiple PR (Proportional Resonance) controller which enhances the output impedance only at the selected frequency, as shown in Figure 13. As mentioned in [18], the repetitive control is equivalent to a combination of a proportional controller and multiple parallel resonant controller. The effectiveness of enhancing output impedance is similar to the multiple PR controller. Meanwhile, the repetitive control is a relative complex for implementation. Therefore, the proposed control strategy with the capacitor feedforward control loop is convenient to enhance the output impedance.

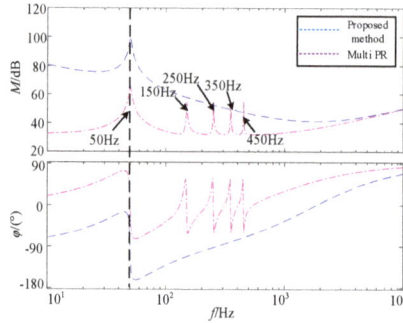

Figure 13. Comparison of the output impedance with the proposed method and multiple PR (Proportional Resonance) controller.

The stability of the proposed control strategy can be determined by the ratio of the grid impedance and output impedance, that is, Z_g/Z'_{ou} should be satisfied with the Nyquist criterion [18]. Here, the grid impedance is 0.2 Ω/0.08 mH, and Z'_{ou} is shown in (11).

It can be found that in Figure 14, there is no pole at the right side of the s plane, and its Nyquist curve does not circle the point $-1 + j0$. According to the Nyquist criterion, this system is stable.

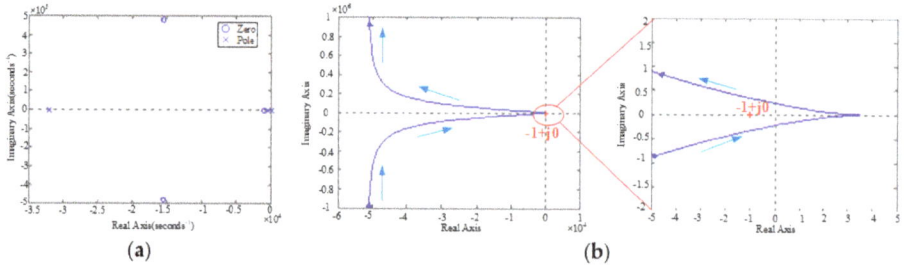

Figure 14. Pole-zero plot and Nyquist curve of Z_g/Z'_{ou}. (a) Pole-zero plot; (b) Nyquist curve, respectively.

5. Simulations and Experimental Verifications

To verify the proposed control strategy, a single-phase grid-tied inverter with an LCL filter was established in PSIM (Power Simulation), and the parameters are shown in Table 1. The 3rd, 5th, 7th, 9th, 11th, and 13th harmonics were injected into grid voltage, and the magnitude of each harmonic was 5%, 6%, 1%, 1.5%, 3.5%, and 3%, respectively.

In order to be more realistic, a 0.5 Ω resistor was concatenated on the DC side to simulate the true DC voltage source in this paper. Figure 15a shows the simulation results of a dual-loop control strategy with capacitor current and grid current. It can be seen that both the voltage and current are heavily distorted. The THD (total harmonic distorted) of the grid current was 7.4%, as shown in Figure 15b. (i_g*20)/A in Figures 15a, 16a and 17a indicates that the grid current is 20 times greater than the original value.

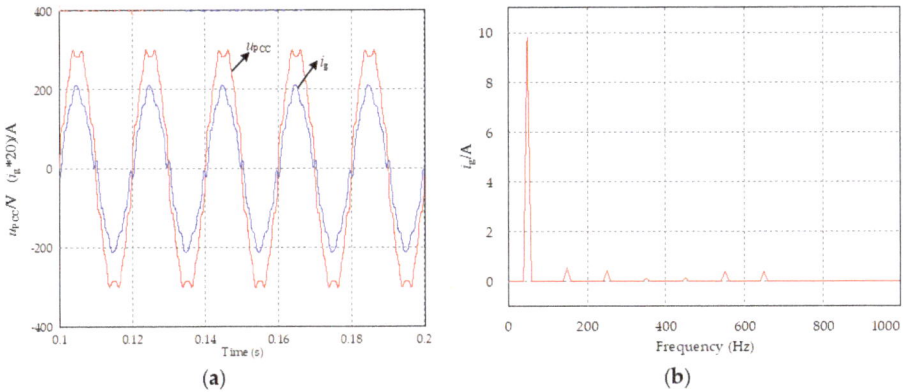

(a)

(b)

Figure 15. Simulation results of dual-loop control strategy. (**a**) Waveforms of grid voltage and current. (**b**) FFT (Fast Fourier Transformation) analysis of grid current (THD = 7.4%).

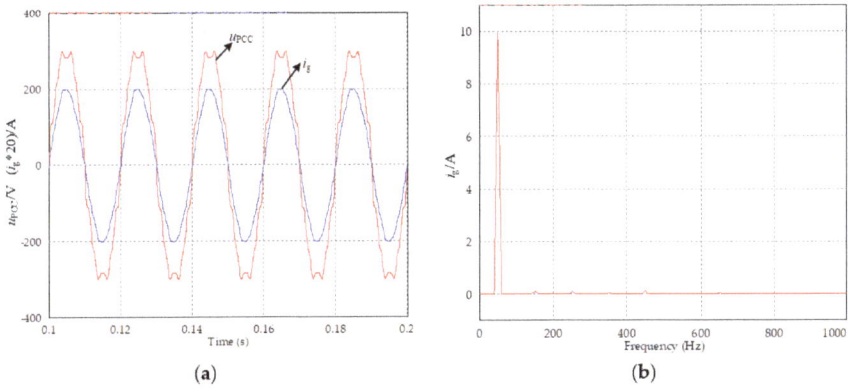

(a)

(b)

Figure 16. Simulation results of grid voltage feedforward control strategy. (**a**) Waveforms of grid voltage and current. (**b**) FFT analysis of grid current (THD = 3.3%).

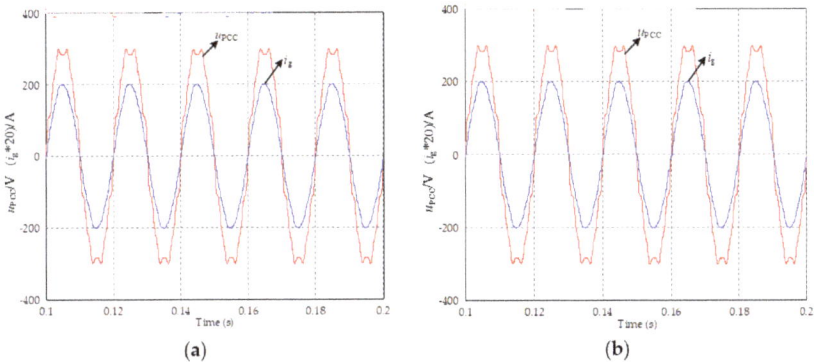

(a)

(b)

Figure 17. Simulation results of the proposed control strategy. (**a**) Waveforms of grid voltage and current. (**b**) FFT analysis of grid current (THD = 2.8%).

By introducing the grid voltage feedforward control loop, the impact of distorted grid voltage on the grid current is smaller. The current wave in Figure 16a is approximately sinusoidal and the THD of the grid current reduces to 3.3% in Figure 16b. It is suggested that by enhancing the output impedance, the current harmonics are suppressed, and the grid voltage feedforward control strategy is effective at suppressing the grid voltage disturbance.

The simulation results in Figure 17a indicate that the proposed control strategy employing the capacitor voltage and grid current is also effective at suppressing current harmonics. It is suggested that the proposed control strategy also has an identical ability of enhancing output impedance with the grid voltage feedforward control strategy. The THD of the grid current is 2.8% in Figure 17b and it achieves the requirements of grid connectedness.

In order to further verify the proposed control strategy, a prototype was designed and established in the laboratory, as shown in Figure 18. The parameters were identical to the simulation as shown in Table 2. The experimental results of the dual-loop control strategy with the capacitor current and grid current feedback, the grid voltage feedforward control, the proposed control strategy of capacitor voltage, and the grid current are shown in Figures 19–21 respectively.

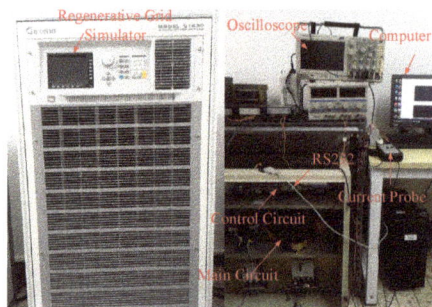

Figure 18. Platform of the grid-tied inverter for single phase.

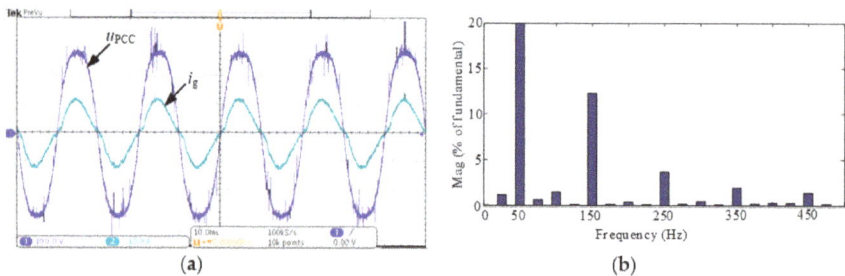

Figure 19. The experimental results of conventional dual-loop control strategy. (**a**) Waveforms of grid voltage and current. (**b**) The FFT analysis of grid current (THD = 13.17%).

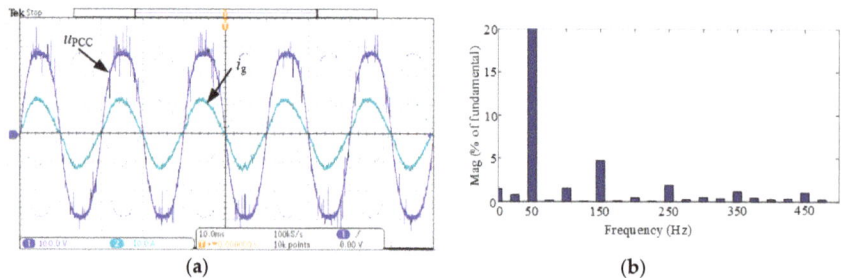

Figure 20. The experimental results of the FF (Feed Forward) strategy. (**a**) Waveforms of grid voltage and current, and (**b**) the FFT analysis of grid current (THD = 5.78%).

48

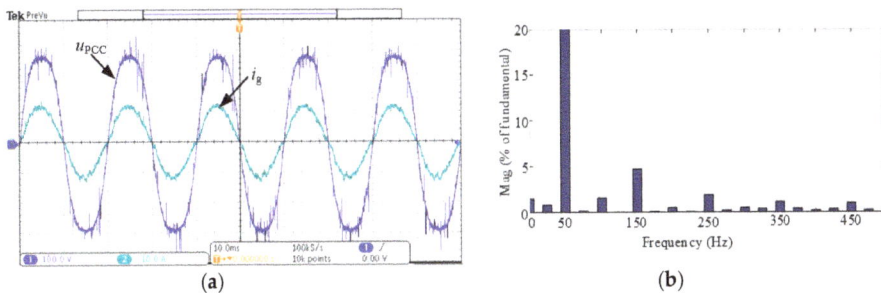

Figure 21. The experimental results of the proposed control strategy. (**a**) Waveforms of grid voltage and current, and (**b**) the FFT analysis of grid current (THD = 5.75%).

The current harmonics are well suppressed when the grid voltage feedforward control is employed, as shown in Figures 19 and 20. The proposed control strategy achieves the same ability of suppressing current harmonics with the FF as shown in Figure 21. The THDs of the grid current with different harmonic suppression schemes are shown in Table 2.

To show the good transient performances, dynamic experiments were carried out in which the reference current was given from 2 to 4 A at t_1 and from 4 to 2 A at t_2, respectively. The experimental results are shown in Figure 22 and indicate that dynamic performance can be guaranteed when the proposed control method is adopted.

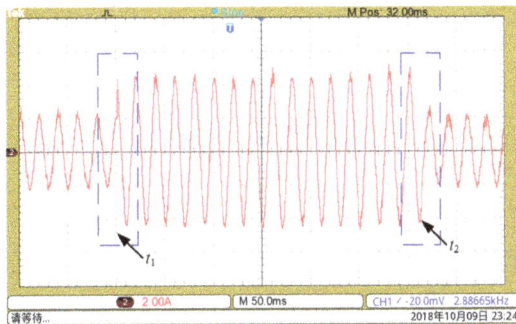

Figure 22. Dynamic performances when the proposed control strategy is employed.

6. Conclusions

The principle and implementation of control-loop-based impedance enhancement of grid-tied inverters for harmonic suppression was proposed in this paper in detail. Based on the idea of reshaping output impedance, two basic methods of impedance reshaping were obtained. It was found that the grid voltage feedback control has an infinite impedance enhancement ability which can eliminate the influences of the distorted grid voltage in theory. A specific example was illustrated to prove the effectiveness of the basic methods of reshaping impedance. Based on the idea of infinite impedance reshaping, the implementation of control loops was made step by step. To find an optimized control loop, a novel control loop that can eliminate the inner capacitor current loop and reduce the number of sensors was proposed through splitting the grid voltage. Finally, the simulation and experimental results were illustrated to verify the effectiveness of the proposed control strategy.

Author Contributions: F.W. and L.Z. provided the original idea of this paper and organized the manuscript. H.G. and X.F. attended the discussions when the simulations and experiments were carried out. All the authors gave the comments and suggestions on the writing and descriptions on the manuscript.

Funding: This work is supported by National Key R&D Program of China (2017YFE0112400) and National Natural Science Foundation of China (51577113).

Conflicts of Interest: The authors declare no conflict of interest.

References

1. Blaabjerg, F.; Teodorescu, T.; Liserre, M.; Timbus, A.V. Overview of control and grid synchronization for distributed power generation systems. *IEEE Trans. Ind. Electron.* **2006**, *51*, 1398–1409. [CrossRef]
2. IEEE Standards Coordinating Committee. *IEEE Standard for Interconnecting Distributed Resources with Electric Power Systems*; IEEE Std 1547-2003; IEEE: Piscataway, NJ, USA, 2018.
3. Xin, Z.; Loh, P.C.; Wang, X.; Blaabjerg, F.; Tang, Y. Highly accurate derivatives for LCL-filtered grid converter with capacitor voltage active damping. *IEEE Trans. Power Electron.* **2016**, *31*, 3612–3625. [CrossRef]
4. Tang, Y.; Loh, P.C.; Wang, P.; Choo, F.H.; Gao, F. Exploring inherent damping characteristic of LCL-filters for three-phase grid-connected voltage source inverters. *IEEE Trans. Power Electron.* **2012**, *27*, 1433–1443. [CrossRef]
5. Chenlei, B.; Xinbo, R.; Xuehua, W.; Weiwei, L.; Donghua, P.; Kailei, W. Step-by-step controller design for LCL-type grid-connected inverter with capacitor–current-feedback active-damping. *IEEE Trans. Power Electron.* **2014**, *29*, 1239–1253. [CrossRef]
6. Wang, F.; Duarte, J.L.; Hendrix, M.A.M. Modeling and analysis of grid harmonic distortion impact of aggregated DG Inverters. *IEEE Trans. Power Electron.* **2011**, *26*, 786–797. [CrossRef]
7. Guoqiao, S.; Xuancai, Z.; Jun, Z.; Dehong, X. A new feedback method for PR current control of LCL-filter-based grid-connected invertir. *IEEE Trans. Ind. Electron.* **2010**, *57*, 2033–2041. [CrossRef]
8. Guoqiao, S.; Dehong, X.; Luping, C.; Xuancai, Z. An improved control strategy for grid-connected voltage source inverters with an LCL filter. *IEEE Trans. Power Electron.* **2008**, *23*, 1899–1906. [CrossRef]
9. Wang, X.; Ruan, X.; Liu, S.; Tse, C.K. Full feedforward of grid voltage for grid-connected inverter with LCL filter to suppress current distortion due to grid voltage harmonics. *IEEE Trans. Power Electron.* **2010**, *25*, 3119–3127. [CrossRef]
10. Li, W.; Ruan, X.; Pan, D.; Wang, X. Full-feedforward schemes of grid voltages for a three-phase LCL-type grid-connected invertir. *IEEE Trans. Ind. Electron.* **2013**, *60*, 2237–2250. [CrossRef]
11. Abeyasekera, T.; Johnson, C.M.; Atkinson, D.J.; Armstrong, M. Suppression of line voltage related distortion in current controlled grid connected inverters. *IEEE Trans. Power Electron.* **2005**, *20*, 1393–1401. [CrossRef]
12. Park, S.Y.; Chen, C.L.; Lai, J.S.; Moon, S.R. Admittance compensation in current loop control for a grid-tie LCL fuel cell inverter. *IEEE Trans. Power Electron.* **2005**, *23*, 1716–1723. [CrossRef]
13. Yang, D.; Ruan, X.; Wu, H. Impedance shaping of the grid-connected inverter with LCL filter to improve its adaptability to the weak grid condition. *IEEE Trans. Power Electron.* **2014**, *29*, 5795–5805. [CrossRef]
14. Xu, J.; Tang, T.; Xie, S. Evaluations of current control in weak grid case for grid-connected LCL-filtered invertir. *IET Power Electron.* **2013**, *6*, 227–234. [CrossRef]
15. Yi, L.; Wei, X.; Chaoxu, M.; Zhengming, Z.; Hongbing, L.; Zhiyong, L. New hybrid damping strategy for grid-connected photovoltaic inverter with LCL filter. *IEEE Trans. Appl. Supercond.* **2014**, *24*. [CrossRef]
16. Pappalardo, C.M.; Guida, D. Use of the Adjoint Method for Controlling the Mechanical Vibrations of Nonlinear Systems. *Machines* **2018**, *6*, 19. [CrossRef]
17. Yang, Y.; Zhou, K.; Wang, H.; Blaabjerg, F.; Wang, D.; Zhang, B. Frequency adaptive selective harmonic control for grid-connected inverters. *IEEE Trans. Power Electron.* **2015**, *30*, 3912–3924. [CrossRef]
18. Sun, J. Impedance-based stability criterion for grid-connected inverters. *IEEE Trans. Power Electron.* **2011**, *26*, 3075–3078. [CrossRef]

energies

MDPI

Article

Modeling and Enhanced Error-Free Current Control Strategy for Inverter with Virtual Resistor Damping

Cheng Nie [1,2], Yue Wang [1,2], Wanjun Lei [1,2,*], Tian Li [2] and Shiyuan Yin [2]

[1]   State Key Laboratory of Electrical Insulation and Power Equipment, Xi'an Jiaotong University, Xi'an 710049, China; niecheng@stu.xjtu.edu.cn (C.N.); yuewang@mail.xjtu.edu.cn (Y.W.)
[2]   Key Laboratory of Shaanxi Smart Grid, Xi'an Jiaotong University, Xi'an 710049, China; litian3117306094@stu.xjtu.edu.cn (T.L.); 12291162@bjtu.edu.cn (S.Y.)
*   Correspondence: leiwanjun@xjtu.edu.cn; Tel.: +86-29-8266-8666

Received: 31 July 2018; Accepted: 17 September 2018; Published: 20 September 2018

Abstract: In microgrid, the grid-connected inverter current with the LCL (inductor-capacitor-inductor) output filter is amplified at certain frequencies. Using virtual resistor damping method can help suppress the amplification. By choosing an appropriate virtual resistor value, the model of the inverter current control loop is simplified as a 2nd-order lowpass filter. Based on such simplified model, this paper proposes a design method of reference current compensation controller, which does not require decomposition of harmonic components. With the reference compensation, the inverter output current control precision is improved obviously. The simulation and experimental results verify the accuracy of the inverter simplified model and effectiveness of the reference compensation design method.

Keywords: microgrid; power quality control; optimal virtual resistor; 2nd-order lowpass filter; reference current compensation

1. Introduction

With the increasing concerns of traditional energy shortage, greenhouse gas emissions, environmental pollution problem and so on, applications of microgrid become popular [1,2]. The typical microgrid system contains traditional power grid and DG (Distributed Generation) system [3–5]. The technology of DG system has been developed rapidly [6,7]. A lot of research results show that the DG inverter has the power quality control ability [8,9], such as being used to compensate load harmonic current [10,11]. For harmonic current compensation, the output harmonic current control performance directly affects the harmonic current compensation effect of DG inverter.

In microgrid, the inverter needs an effective filter [12,13] to avoid switching and multiple-switching frequency harmonics [14]. Compared with the L filter, the LCL filter is a more attractive solution [15,16]. However, because the potential resonance problem [17–20], precise current control of inverter with LCL filter is difficult in practical application.

The resonance problem can be solved by active damping method. In general, active damping is realized by introducing one more feedforward variable into the control loop [21]. There are mainly two situations. First, for grid-side current control, feeding back converter-side current [22] or capacitor current [23,24] can both achieve good damping effect. In [22], the current control loop is double-loop structure. The outer loop controls the output current and the inner loop stabilizes the system. In [23,24], the feeding back capacitor current is used to generate an additional current reference. This is equivalent to adding a resistor to the mathematical model. They only focus on the control of fundamental component. The harmonic component control with virtual resistor damping is not analyzed [25–28]. Second, for inverter-side current control, capacitor voltage contributes to stabilize the system [29–31]. In [30], dead-beat control is used as current controller and the capacitor harmonic voltage is introduced into the current control loop through a proportional controller, suppressing resonance effectively. But the control target is only the output fundamental current, the effect on the output harmonic current control is also not analyzed. In [31], based on NORTON's equivalence circuit, the effect of virtual resistance on the inverter output harmonic current amplitude is studied. The output harmonic current phase delay is compensated by closed-loop control of filter capacitor harmonic voltage phase. Although the output harmonic current is controlled, the out harmonic current phase is not sampled and the capacitor voltage phases of each harmonics are extracted separately. The compensation is indirect and discrete.

The purpose of this paper is to propose a simple control method to improve the effect of inverter harmonic current control. Firstly, the equivalent current source model of inverter with virtual resistor is established. Secondly, the paper has analyzed the relationship between virtual resistor value and inverter harmonic current control character. The model of the inverter current control loop can be simplified as 2nd-order lowpass filter by choosing appropriate resistor value, which can be obtained according to the quality factor regulation of the simplified model. Thirdly, the phase lag angle of inverter output harmonic current and the harmonic compensation error are analyzed quantitatively. Then the design method of reference compensation controller is proposed based on the simplified model. And then, the paper discusses the influence of LCL parameters on reference compensation control. Finally, the simulation and experimental results show the correctness of the simplified model and the effectiveness of the reference compensation.

2. Modelling of the Inverter with Virtual Resistor Control

The configuration of system with grid-connected distributed inverter of the microgrid is shown in the Figure 1. There is power grid, PV and its inverter. They provide electricity to the load together.

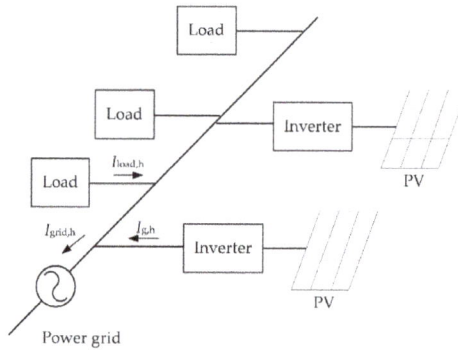

Figure 1. Circuit diagram of microgrid.

In the microgrid system, except providing fundamental energy, distributed inverter can also output harmonic current $I_{g,h}$ to compensate the load harmonic current $I_{load,h}$. There are already lots of research on the distributed inverter control and mainly focus on fundamental component [32–34]. In this section, harmonic current control model of the inverter with virtual resistor damping is established.

The schematic diagram of distributed inverter is presented in Figure 2, the upper part is the main circuit configuration and the lower part is the digital control diagram. L_1 is the converter-side filter inductance and L_2 is the grid-side filter inductance, C_f is the filter capacitor.

The current control method here is based on the traditional proportional control method. Converter side current I_L is selected as the controlled object and filter capacitor voltage V_C is measured for system synchronization and active damping, in this way only two feedback sensors are needed. The inverter output voltage is expressed as Equation (1):

$$V_{PWM}(s) = V_C(s) + K_p(I_{ref}(s) - I_{VD}(s) - I_L(s)) \tag{1}$$

Here $I_{ref}(s)$ and $I_L(s)$ are the reference current and the measured converter-side feedback current, respectively. K_p is a parameter of current controller, equals to a constant value. $V_c(s)$ is the measured capacitor voltage, used to compensate the disturbance of ac side voltage. $I_{VD}(s)$ is the damping current, which is generated by capacitor harmonic voltage, as shown in Equation (2).

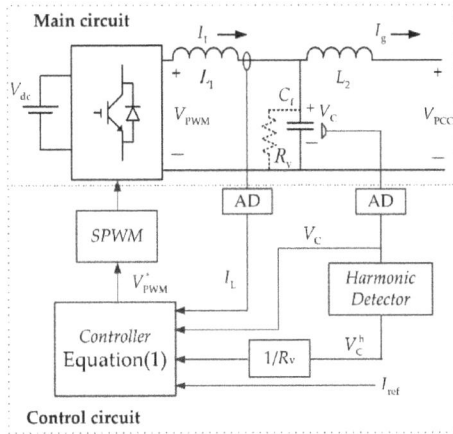

Figure 2. Conventional proportion-controlled DG system with virtual damping.

$$I_{VD}(s) = \frac{V_C(s)}{R_V} \tag{2}$$

In the Equation (2), R_V is a constant value with the unit ohms, equals to the virtual resistor's value. In this way, the virtual resistor is parallel connect with the filter capacitor, as in Figure 2. Based on the open-loop transfer function of LCL network, Equations (3)–(5), the relationship between the output current $I_g(s)$ and the reference current $I_{ref}(s)$ of the inverter can be obtained, as Equation (6):

$$I_L(s) = H_1(s)V_{PWM}(s) + H_2(s)V_{PCC}(s) \tag{3}$$

$$I_g(s) = H_3(s)V_{PWM}(s) + H_4(s)V_{PCC}(s) \tag{4}$$

$$V_C(s) = H_5(s)V_{PWM}(s) + H_6(s)V_{PCC}(s) \tag{5}$$

The coefficients $H_1(s)$–$H_6(s)$ are determined by LCL filter parameters and are shown at the end of this section.

$$I_g(s) = G_T^{ad}(s)I_{ref}(s) - Y_{eq}^{ad}(s)V_{PCC}(s) \tag{6}$$

The coefficients $G_T^{ad}(s)$ and $-Y_{eq}^{ad}(s)$ describe the output current responses to the current reference and the PCC voltage, respectively. The detailed expressions are described as follows:

$$G_T^{ad}(s) = \frac{K_p H_3(s)}{1 - H_5(s)(1 - \frac{K_p}{R_v}) + K_p H_1(s)} \tag{7}$$

$$Y_{eq}^{ad}(s) = \frac{K_p H_2(s)H_3(s) - H_3(s)H_6(s)(1 - \frac{K_p}{R_v})}{1 - H_5(s)(1 - \frac{K_p}{R_v}) + K_p H_1(s)} - H_4(s) \tag{8}$$

Figure 3 describes the corresponding Norton's equivalent circuit of the inverter. $G_T^{ad}(s)$ behaves as the coefficient of current source and $Y_{eq}^{ad}(s)$ represents as the associated parallel admittance.

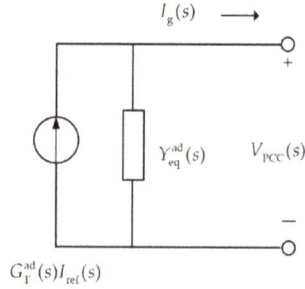

Figure 3. Equivalent current source model of the inverter.

$$H_1(s) = \frac{L_2 C_f s^2 + 1}{L_1 L_2 C_f s^3 + (L_1 + L_2)s} \tag{9}$$

$$H_2(s) = -\frac{1}{L_1 L_2 C_f s^3 + (L_1 + L_2)s} \tag{10}$$

$$H_3(s) = \frac{1}{L_1 L_2 C_f s^3 + (L_1 + L_2)s} \tag{11}$$

$$H_4(s) = -\frac{L_1 C_f s + 1}{L_1 L_2 C_f s^3 + (L_1 + L_2)s} \tag{12}$$

$$H_5(s) = \frac{L_2 s}{L_1 L_2 C_f s^3 + (L_1 + L_2)s} \tag{13}$$

$$H_6(s) = \frac{L_1 s}{L_1 L_2 C_f s^3 + (L_1 + L_2)s} \tag{14}$$

3. Optimal Virtual Resistor Value of the Inverter

Form the Equation (7), the value of virtual resistor affects the coefficients $G_T^{ad}(s)$. It means R_v value can regulate the harmonic current control performance. Bode plot of coefficient $G_T^{ad}(s)$ with different R_v is shown in Figure 4. Table 1 lists the parameters of the system.

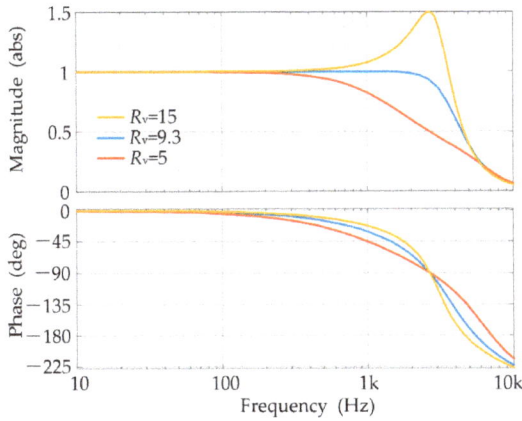

Figure 4. Bode plot of $G_T^{ad}(s)$ with different virtual resistor.

Table 1. Parameters of the System.

Symbol	Value	Symbol	Value
L_1	0.6 mH	V_{dc}	600 V
L_2	0.6 mH	V_{grid}	380 V (50 Hz)
C_f	6 μF	K_p	30
L_{grid}	0.1 mH	f_s	20 kHz

From Figure 4, the R_v value has both effect on the amplitude-frequency and phase-frequency characteristics of $G_T^{ad}(s)$. From the amplitude-frequency characteristic curve, as the R_v value decreases, the curve decreases more near the resonant frequency. And there is an optimal virtual resistance value (R_v = 9.3 Ω), which can make the amplitude-frequency characteristic curve closer to unit 1 below 2 kHz. From the phase-frequency characteristic curve, all the curves cross 90° at resonance frequency. Below the resonance frequency, smaller R_v value responses to more phase lag; above the resonance frequency, smaller R_v value response to less phase lag.

The coefficients $G_T^{ad}(s)$ is 3-order s-domain function, related to the LCL parameters, current control loop K_p and virtual resistance R_v, as follow:

$$G_T^{ad}(s) = \frac{K_p}{L_1 L_2 C s^3 + K_p L_2 C s^2 + (L_1 + K_p L_2 / R_v)s + K_p} \tag{15}$$

In Equation (15), the coefficient of s^3 term is 2.16×10^{-12}, and the coefficient of s^2 is 1.08×10^{-7}, which is about 10^5 times of the coefficient of s^3. When the s^3 term in Equation (15) is removed, the approximate expression of $G_T^{ad}(s)$ is obtained, as Equation (16):

$$\hat{G}_T^{ad}(s) = \frac{1/(L_2 C_f)}{s^2 + (L_1 + K_p L_2 / R_v)/(K_p L_2 C_f)s + 1/(L_2 C_f)} \tag{16}$$

When Equation (16) is regarded as the expression of 2nd-order lowpass filter, the corresponding cutoff frequency and quality factor are respectively.

$$\omega_n = \sqrt{\frac{1}{L_2 C_f}} \tag{17}$$

$$\delta = \frac{L_1 R_v + K_p L_2}{2 K_p R_v} \omega_n \tag{18}$$

According to Equation (17), the cutoff frequency is a function of the grid-side inductor L_2 and the filter capacitor C_f. The quality factor is a function of the virtual resistor R_v and LCL parameters. For 2nd-order lowpass filter, when the value of quality factor is 0.707, the amplitude-frequency characteristic curve has not overshoot. According to the parameters in Table 1, the cutoff frequency is 16,666.67 rad/s, and the optimal virtual resistor values is 9.3 Ω.

Figure 5 shows a comparison between the transfer function $G_T^{ad}(s)$ and the transfer function of the second order low pass filter (LPF). In Figure 5, the model of the inverter with optimal virtual resistor damping (blue line) is very similar to that of the 2nd-order lowpass filter (red line). For the amplitude-frequency characteristic, the inverter and the 2nd-order lowpass filter are basically the same below 1 kHz. Then, as the frequency increases, the attenuation of the 2nd-order lowpass filter is greater than that of the inverter. For the phase-frequency characteristic, the two models almost coincide with each other below resonant frequency.

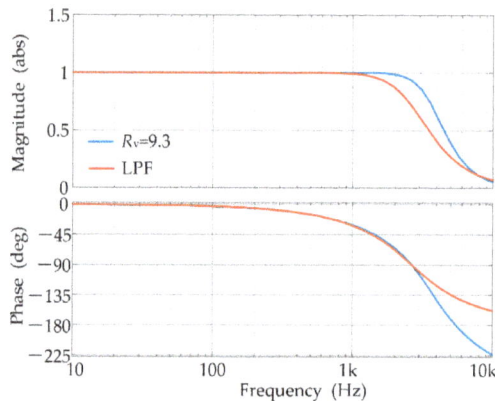

Figure 5. Comparison between the inverter and LPF.

4. Reference Current Compensation Method

With the optimal virtual resistor damping, the output current control performance of the inverter can be approximately a 2nd-order lowpass filtering model. Although the amplitude of the output current is almost the same as the reference current, but the phase lag problem needs be considered carefully. At the frequency 1.5 kHz, the phase lag degree is almost 45 degrees. This will affect the harmonic current compensation ability of inverter.

The relationship of current vector of system in Figure 1 is shown in Figure 6.

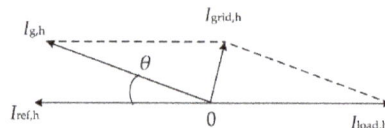

Figure 6. Diagram of current vector relationship.

In Figure 6, $I_{load,h}$ is the load harmonic current. Naturally, the amplitude-equal and phase-inverse vector I_{ref_h} is the reference harmonic current of the inverter. As mentioned before, the output current is amplitude-equal but phase-lag, as vector $I_{g,h}$ in Figure 6. The synthetic vector $I_{grid,h}$ is the grid harmonic current. The fact that $I_{grid,h}$ is not equal to zero means poor harmonic current compensation performance of the inverter.

In order to evaluate the influence of phase lag on current control, harmonic compensation error rate α is defined, as shown in Equation (19).

$$\alpha = \frac{\left|I_{grid,h}\right|}{\left|I_{load,h}\right|} \tag{19}$$

In Equation (19), the numerator is the grid harmonic current, the denominator is the load harmonic current. When without any compensation, the grid harmonic current is equal to the load harmonic current, the ratio α is equal to unit 1. When with compensation, the grid harmonic current is less than the load harmonic current, the ratio α is less than unit 1. When the α is equal to 0, means no harmonic component in the grid current, also corresponds to the best current control effect.

With optimal virtual resistor, the amplitude of the output current is the same as the reference current. Then the harmonic compensation error rate is a function of lag angle, as Equation (20):

$$\alpha = \frac{\sqrt{\left|I_{g,h}\right|2 + \left|I_{Load,h}\right|2 - 2\left|I_{g,h}\right|\left|I_{Load,h}\right|\cos\theta}}{\left|I_{Load,h}\right|} \times 100\% \tag{20}$$

Table 2 shows the phase delay angle and compensation error rate below f < 1.5 kHz. From Table 2, phase lag problem will lead to serious harmonic compensation error, especially for the high frequency harmonics. Therefore, it is necessary to compensate the phase of the reference current to reduce the current control error.

Table 2. Phase Delay and Harmonic Compensation Error.

Harmonic Order	5th	7th	11th	13th	17th	19th	23th	25th	29th
θ	7.6°	10.7°	16.8°	19.9°	26.2°	29.4°	35.7°	38.9°	45.6°
α (%)	1.2	2.5	6.0	8.4	14.5	18.2	26.6	31.4	42.5

Figure 7 shows the principle of reference compensation control.

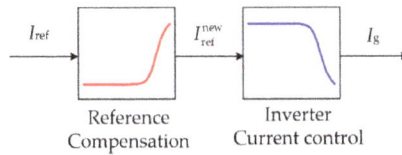

Figure 7. Schematic diagram of reference compensation.

The reference current I_{ref} first goes through a regulation loop, whose function is to generate a new reference I_{ref}^{new}. The new reference is phase ahead of the old ones, and the leading angle increases with the frequency increase. Although the output current is still lagging the new reference I_{ref}^{new}, but the output current is consistent with reference current I_{ref}. In this way, the phase delay of the output current is compensated.

Considering that the current control model of the inverter with optimal virtual resistor is basically agree with the 2nd-order lowpass filter in the phase-frequency characteristics. Therefore, the reference current regulation method proposed in this paper is based on the inverter's 2nd-order lowpass filtering model $G_T^{ad}(s)$.

The reference compensation process is to divide the reference current $I_{ref}(s)$ by the inverter's 2nd-order lowpass filter model, as Equation (21).

$$I_{ref}^{new}(s) = I_{ref}(s) \cdot \frac{1}{\hat{G}_T^{ad}(s)} \tag{21}$$

With the reference compensation, the inverter output current is:

$$I_g^{ad}(s) = I_{ref}(s)G(s) - Y_{eq}^{ad}(s)V_{PCC}(s) \tag{22}$$

Here,

$$G(s) = G_T^{ad}(s) \cdot \frac{1}{\hat{G}_T^{ad}(s)} \tag{23}$$

The characteristics of $G(s)$ determine the inverter current control character. Figure 8 is the bode plot of $G(s)$. It can be seen from the amplitude-frequency characteristic of the coefficient $G(s)$ that below 1 kHz, the curve is basically equal to unit 1. From the phase-frequency characteristic, the phase lag problem is well compensated below 2.5 kHz. Above all, for the application of frequency f < 1.5 kHz, the reference current compensation based on 2nd-order lowpass filter model has obvious effect.

The comparison results of harmonic compensation error rate without and with reference compensation method are shown in Table 3.

As can be seen from the data in Table 3, when there is no reference compensation, the error rate of harmonic compensation is large, and the compensation error rate increases with the increase of harmonic frequency. While with the reference compensation, the compensation error rate of each harmonic has been significantly reduced.

Table 3. Error of the harmonic current compensation.

Control Mode	$\alpha/\%$								
	5th	7th	11th	13th	17th	19th	23th	25th	29th
Without reference compensation	1.2	2.5	6.0	8.4	14.5	18.2	26.6	31.4	42.5
With reference compensation	0.1	0.2	0.4	0.6	1.3	1.5	2.6	3.5	5.6

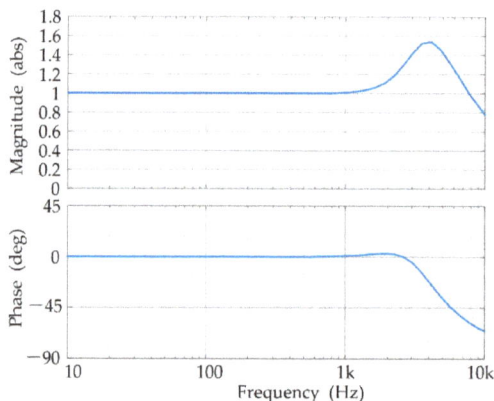

Figure 8. Bode plot of the coefficient $G(s)$.

5. Effect of LCL Parameters on Performance of the Reference Compensation Method

The 2nd-order lowpass filter model of the inverter ignores the three orders item of the transfer function $G_T^{ad}(s)$ in its closed-loop model, which affects the final reference compensation effect. As can be seen from Figure 8, when the frequency is less than 1.5 kHz, the phase lags problem is well compensated. But with the frequency increasing, the amplitude frequency-characteristic of $G(s)$ has overshoot. For different LCL parameters, the effect of reference compensation control is different.

Many literatures have studied the parameter design methods of LCL filter. The LCL parameter design method is not studied in this paper. Based on the parameter design method [33,34], this paper selects four different LCL parameters, as shown in Table 4. For each group, the grid-side inductance is equal to the inverter-side inductance. And they have the same resonance frequency.

Table 4. LCL parameters for comparison.

Symbol	$L_1 = L_2$	C_f	f_r	f_{LPF}
1	0.8 mH	4.5 μF		
2	0.6 mH	6 μF	3700 Hz	2653 Hz
3	0.4 mH	9 μF		
4	0.2 mH	18 μF		

In Equation (24) is the resonant frequency of LCL filter.

$$f_r = \frac{1}{2\pi}\sqrt{\frac{L_1 + L_2}{L_1 L_2 C}} = \frac{1}{2\pi}\sqrt{\frac{2}{L_2 C}} = \sqrt{2} f_{LPF} \tag{24}$$

Equation (24) shows that when $L_1 = L_2$ and the resonance frequency is the same, the four groups of parameters have the same 2nd-order lowpass filter model. The bode plot of $G(s)$ is shown in the Figure 9.

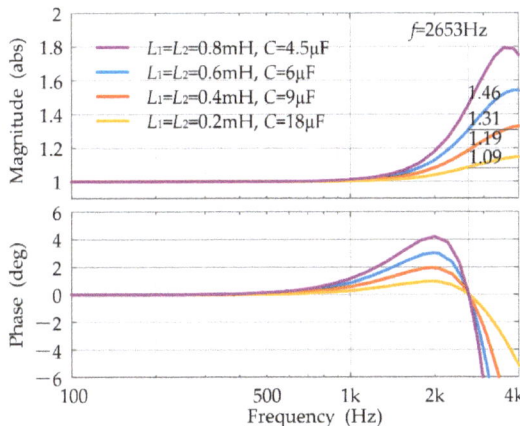

Figure 9. Bode plot of $G(s)$ with different LCL parameters.

In the Figure 9, the four phase-frequency characteristic curves crossing 0 degree at the same frequency. The simplified expression of $G(s)$ is as follows:

$$G(s) = \frac{s^2 + 1.414\omega_c s + \omega_c^2}{\frac{L_1}{K_p}s^3 + s^2 + 1.414\omega_c s + \omega_c^2} \tag{25}$$

The zero-cross frequency of the phase-frequency characteristic curves can be obtained by solving Equation (26).

$$\angle|G(s)| = 0° \tag{26}$$

The result is $\omega_x = \omega_{LPF}$, it is just equal to cut-off frequency of the inverter's 2nd-order lowpass filter model. And the amplitude value at this frequency is:

$$|G(s)| = \frac{1.414}{1.414 - L_1\omega_{LPF}/K_p} \tag{27}$$

According to the above analysis, LCL parameters have the following two effects on the reference compensation control:

1. Below cut-off frequency, the higher the inductance value, the greater the overshoot of the amplitude-frequency curve; Conversely, the lower the inductance value, the amplitude-frequency curve is closer to unit 1.
2. Below cut-off frequency, the higher the inductance value, the greater the overshoot of the phase-frequency curve; Conversely, the lower the inductance value, the phase-frequency curve is closer to zero.

In conclusion, the smaller the inductance value, the better the effect of reference compensation.

6. Simulation and Experimental Verification

To verify the correctness of the theory and the validity of the method, the theoretical derivation was verified under simulation (PSIM) and experiment (laboratory platform). The proposed current control method is applied to active power filter. Parameters are same as list in Table 1.

6.1. Simulation

In simulation, a controlled current source is used to inject harmonics into the power grid to simulate the nonlinear load. The harmonic order is 5, 11, 17, 23 and 29 (6n − 1). With optimal virtual resistor, the simulation results are shown in Figure 10. From the simulation results, harmonic component in grid current is very large. By comparing the waveform of reference current and output current, the output current has serious phase lag problem.

The simulation results in Figure 11 are with the reference compensation control. The load does not change, so the load current with reference compensation is just same as Figure 10b. In Figure 11, the load harmonic current is the reference current before compensation (Figure 11b). It is the phase-lag of the reference current after compensation (Figure 11c). Because of the "lowpass filter characteristics" of the inverter, the output current (Figure 11d) is also phase-lag of the reference current after compensation (Figure 11c). That makes the output current (Figure 11c) just in accordance with the load harmonic current (Figure 11b). The harmonic component in grid current is reduced obviously. The control effect of inverter output current is guaranteed.

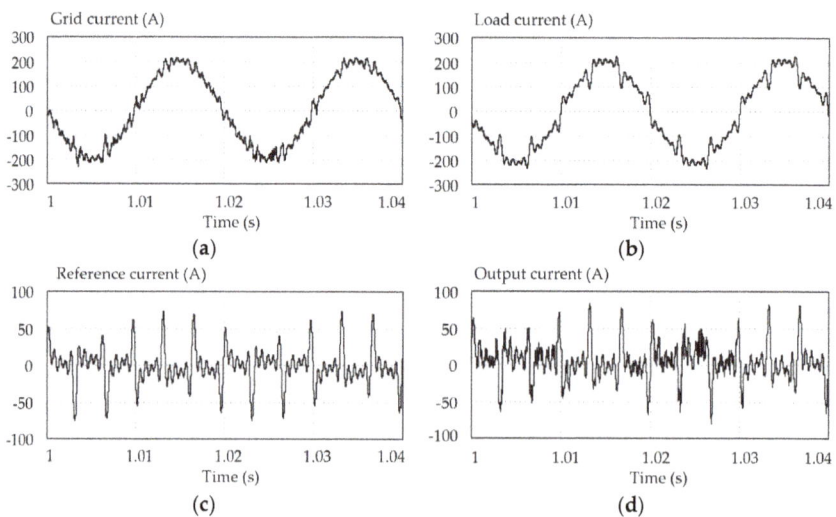

Figure 10. Simulation result without reference compensation. (**a**) grid current; (**b**) load current; (**c**) reference current; (**d**) output current.

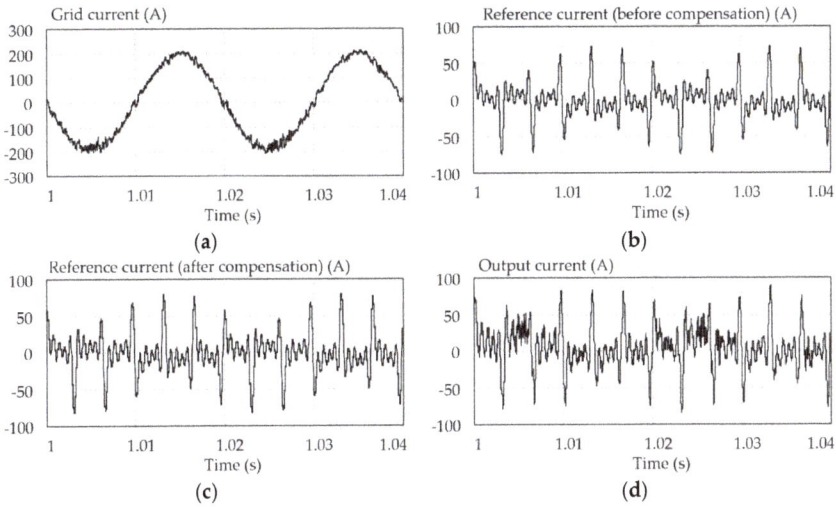

Figure 11. Simulation result with reference compensation. (**a**) grid current; (**b**) reference current (before compensation); (**c**) reference current (after compensation); (**d**) output current.

For different LCL parameters (first set of parameters in Table 4), simulation verification is also carried out. The FFT result of the grid current is shown in Figure 12.

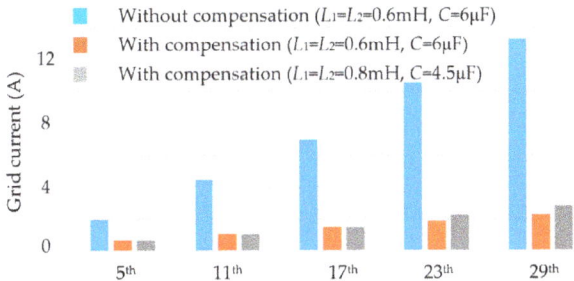

Figure 12. FFT results of grid current.

From Figure 12, when without reference compensation control, the current in power grid contains a large harmonic component, and the harmonic component gradually increases with the increase of frequency. While, with the reference compensation control, the harmonic component in the power grid is obviously smaller, and the improvement effect is still good when the frequency increase. For different LCL parameters, the smaller L value can realize better current control effect.

6.2. Experiment

The experimental setup is shown in Figure 13.

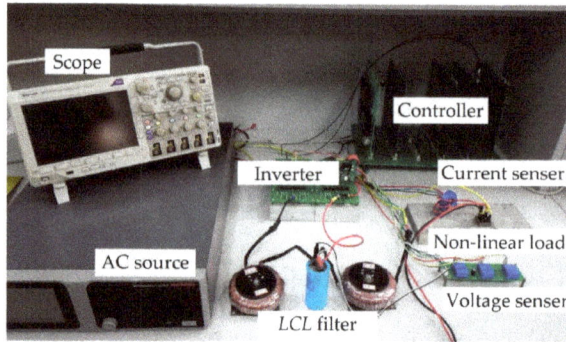

Figure 13. Experimental setup.

The inverter is single-phase 4.5 kVA, four discrete insulated-gate bipolar transistors (Infineon IKW40T120) were employed. DSP (Digital Signal Processor, TI TMS320F2812) is the micro-controller, switching frequency and sampling frequency are both 20 kHz. The nonlinear load is diode rectifier with resistance-inductance load (R_{load} = 40 Ω, L_{load} = 2 mH).

Figure 14 is the experimental result of grid current, load current and inverter output current. Figure 15 is the FFT analysis of the experimental results.

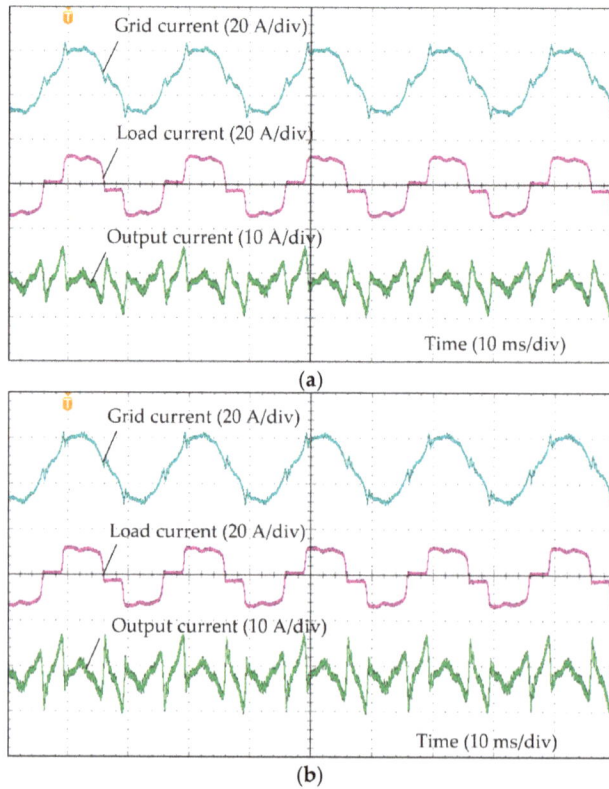

Figure 14. Experimental result. (**a**) Without reference compensation; (**b**) with reference compensation.

Figure 15a is the amplitude of the inverter output current, and Figure 15b is the phase of the inverter output current. For both control method, the amplitude of the inverter output current is almost the same, means the reference compensation control does not change the amplitude of the output current. But, when without reference compensation, the phase delay problem is serious. And the phase lag angle increases as the frequency increases. With the reference compensation control, the phase delay of output current for each harmonic is compensated well.

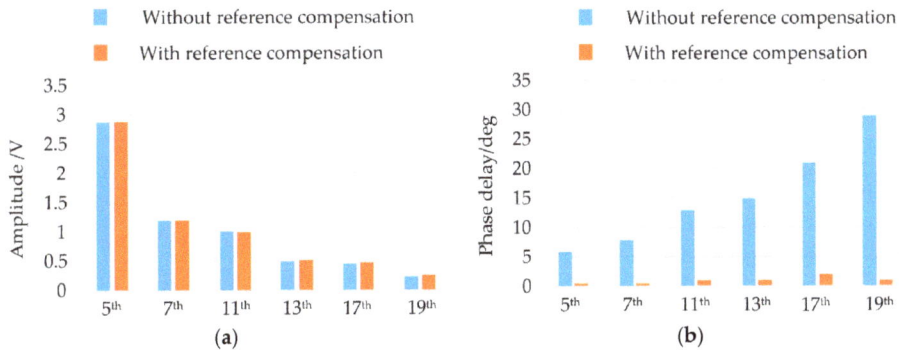

Figure 15. FFT analysis of Experimental result. (**a**) Amplitude; (**b**) phase delay.

7. Conclusions

Virtual resistor damping affects the current control character of the inverter. With optimal virtual resistor value, the output current amplitude control precision is enhanced. In this condition, the current source model of the inverter can be simplified as 2nd-order lowpass filter. It means the simplified model can also be used to design the reference compensation controller. For the proposed reference compensation method, there is no need to decompose harmonic components, and only requires LCL parameters and current control loop parameters. It greatly simplifies the design of reference compensation.

In this paper, only the steady-state current control effect is analyzed. The study of dynamic characteristics, and typical operations are also very necessary and will be carried out in the next step.

Author Contributions: Conceptualization, C.N. and W.L.; Methodology, C.N. and W.L.; Software, C.N. and S.Y.; Validation, C.N., S.Y. and T.L.; Formal Analysis, C.N. and T.L; Investigation, C.N. and W.L.; Resources, W.L. and Y.W.; Data Curation, C.N. and T.L.; Writing—Original Draft Preparation, C.N.; Writing—Review & Editing, C.N.; Visualization, C.N.; Supervision, W.L. and Y.W.; Project Administration, W.L.; Funding Acquisition, W.L. and Y.W. All the authors approved the publication.

Funding: This research was funded by National Natural Science Foundation of China, grant number 51207126.

Conflicts of Interest: The authors declare no conflict of interest.

References

1. Lu, Z.; Li, H.; Qiao, Y. Probabilistic Flexibility Evaluation for Power System Planning Considering Its Association with Renewable Power Curtailment. *IEEE Trans. Power Syst.* **2018**, *33*, 3285–3295. [CrossRef]
2. Nejabatkhah, F.; Li, Y.W. Overview of Power Management Strategies of Hybrid AC/DC Microgrid. *IEEE Trans. Power Electron.* **2015**, *30*, 7072–7089. [CrossRef]
3. Oureilidis, K.O.; Demoulias, C.S. An enhanced role for an energy storage system in a microgrid with converter-interfaced sources. *J. Eng.* **2014**, *11*, 618–625. [CrossRef]
4. Hossain, M.A.; Pota, H.R.; Issa, W.; Hossain, M.J. Overview of AC Microgrid Controls with Inverter-Interfaced Generations. *Energies* **2017**, *10*, 1300. [CrossRef]
5. Kallamadi, M.; Sarkar, V. Generalised analytical framework for the stability studies of an AC microgrid. *J. Eng.* **2016**, *6*, 171–179. [CrossRef]

6.   Li, Y.; Wu, M.; Li, Z. A Real Options Analysis for Renewable Energy Investment Decisions under China Carbon Trading Market. *Energies* **2018**, *11*, 1817. [CrossRef]

7.   Agrawal, R.; Jain, S. Multilevel inverter for interfacing renewable energy sources with low/medium- and high-voltage grids. *IET Renew. Power Gener.* **2017**, *11*, 1822–1831. [CrossRef]

8.   Han, Y.; Li, H.; Shen, P.; Coelho, E.A.; Guerrero, J.M. Review of Active and Reactive Power Sharing Strategies in Hierarchical Controlled Microgrids. *IEEE Trans. Power Electron.* **2017**, *32*, 2427–2451. [CrossRef]

9.   Chilipi, R.R.; Al Sayari, N.; Beig, A.R.; Al Hosani, K. A Multitasking Control Algorithm for Grid-Connected Inverters in Distributed Generation Applications Using Adaptive Noise Cancellation Filters. *IEEE Trans. Energy Convers.* **2016**, *31*, 714–727. [CrossRef]

10.   He, J.; Li, Y.W.; Blaabjerg, F.; Wang, X. Active Harmonic Filtering Using Current-Controlled, Grid-Connected DG Units With Closed-Loop Power Control. *IEEE Trans. Power Electron.* **2014**, *29*, 642–653. [CrossRef]

11.   He, J.; Li, Y.W.; Munir, M.S. A Flexible Harmonic Control Approach through Voltage-Controlled DG–Grid Interfacing Converters. *IEEE Trans. Ind. Electron.* **2012**, *59*, 444–455. [CrossRef]

12.   Liu, Y.; Lai, C.-M. LCL Filter Design with EMI Noise Consideration for Grid-Connected Inverter. *Energies* **2018**, *11*, 1646. [CrossRef]

13.   Xu, J.; Xie, S.; Huang, L.; Ji, L. Design of LCL-filter considering the control impact for grid-connected inverter with one current feedback only. *IET Power Electron.* **2017**, *10*, 1324–1332. [CrossRef]

14.   Hren, A.; Mihalič, F. An Improved SPWM-Based Control with Over-Modulation Strategy of the Third Harmonic Elimination for a Single-Phase Inverter. *Energies* **2018**, *11*, 881. [CrossRef]

15.   He, J.; Li, Y.W. Hybrid voltage and current control approach for DG grid interfacing converters with LCL filters. *IEEE Trans. Ind. Electron.* **2013**, *60*, 1797–1809. [CrossRef]

16.   Jalili, K.; Bernet, S. Design of LCL filters of active-front-end two level voltage-source converters. *IEEE Trans. Ind. Electron.* **2009**, *56*, 1674–1689. [CrossRef]

17.   Tang, Y.; Yao, W.; Loh, P.C.; Blaabjerg, F. Design of LCL Filters with LCL Resonance Frequencies beyond the Nyquist Frequency for Grid-Connected Converters. *IEEE J. Emerg. Sel. Top. Power Electron.* **2016**, *4*, 3–14. [CrossRef]

18.   Tang, Y.; Loh, P.C.; Wang, P.; Choo, F.H.; Gao, F.; Blaabjerg, F. Generalized design of high performance shunt active power filter with output LCL filter. *IEEE Trans. Ind. Electron.* **2012**, *59*, 1443–1452. [CrossRef]

19.   Said-Romdhane, M.B.; Naouar, M.W.; Belkhodja, I.S.; Monmasson, E. An Improved LCL Filter Design in Order to Ensure Stability without Damping and Despite Large Grid Impedance Variations. *Energies* **2017**, *10*, 336. [CrossRef]

20.   Wu, T.F.; Misra, M.; Lin, L.C.; Hsu, C.W. An Improved Resonant Frequency Based Systematic LCL Filter Design Method for Grid-Connected Inverter. *IEEE Trans. Ind. Electron.* **2017**, *64*, 6412–6421. [CrossRef]

21.   Wu, W.; Liu, Y.; He, Y.; Chung, H.S.; Liserre, M.; Blaabjerg, F. Damping Methods for Resonances Caused by LCL-Filter-Based Current-Controlled Grid-Tied Power Inverters: An Overview. *IEEE Trans. Ind. Electron.* **2017**, *64*, 7402–7413. [CrossRef]

22.   Tang, Y.; Loh, P.C.; Wang, P.; Choo, F.H.; Gao, F. Exploring Inherent Damping Characteristic of LCL-Filters for Three-Phase Grid-Connected Voltage Source Inverters. *IEEE Trans. Power Electron.* **2012**, *27*, 1433–1443. [CrossRef]

23.   Liu, F.; Zhou, Y.; Duan, S.; Yin, J.; Liu, B.; Liu, F. Parameter design of a two-current-loop controller used in a grid-connected inverter system with LCL filter. *IEEE Trans. Ind. Electron.* **2009**, *56*, 4483–4491. [CrossRef]

24.   Lorzadeh, I.; Askarian Abyaneh, H.; Savaghebi, M.; Bakhshai, A.; Guerrero, J.M. Capacitor Current Feedback-Based Active Resonance Damping Strategies for Digitally-Controlled Inductive-Capacitive-Inductive-Filtered Grid-Connected Inverters. *Energies* **2016**, *9*, 642. [CrossRef]

25.   Jin, W.; Li, Y.; Sun, G.; Bu, L. H∞ Repetitive Control Based on Active Damping with Reduced Computation Delay for LCL-Type Grid-Connected Inverters. *Energies* **2017**, *10*, 586. [CrossRef]

26.   Yao, W.; Yang, Y.; Zhang, X.; Blaabjerg, F.; Loh, P.C. Design and Analysis of Robust Active Damping for LCL Filters Using Digital Notch Filters. *IEEE Trans. Power Electron.* **2017**, *32*, 2360–2375. [CrossRef]

27.   Chen, C.; Xiong, J.; Wan, Z.; Lei, J.; Zhang, K. A Time Delay Compensation Method Based on Area Equivalence for Active Damping of an LCL-Type Converter. *IEEE Trans. Power Electron.* **2017**, *32*, 762–772. [CrossRef]

28.   Pan, D.; Ruan, X.; Wang, X. Direct Realization of Digital Differentiators in Discrete Domain for Active Damping of LCL-Type Grid-Connected Inverter. *IEEE Trans. Power Electron.* **2018**, *33*, 8461–8473. [CrossRef]

29. Zhou, S.; Zou, X.; Zhu, D.; Tong, L.; Kang, Y. Improved Capacitor Voltage Feedforward for Three-Phase LCL-Type Grid-Connected Converter to Suppress Start-Up Inrush Current. *Energies* **2017**, *10*, 713. [CrossRef]
30. He, J.; Li, Y.W.; Bosnjak, D.; Harris, B. Investigation and Active Damping of Multiple Resonances in a Parallel-Inverter-Based Microgrid. *IEEE Trans. Power Electron.* **2013**, *28*, 234–246. [CrossRef]
31. Nie, C.; Wang, Y.; Lei, W.; Chen, M.; Zhang, Y. An Enhanced Control Strategy for Multiparalleled Grid-Connected Single-Phase Converters with Load Harmonic Current Compensation Capability. *IEEE Trans. Ind. Electron.* **2018**, *65*, 5623–5633. [CrossRef]
32. Liu, J.; Miura, Y.; Ise, T. Comparison of Dynamic Characteristics between Virtual Synchronous Generator and Droop Control in Inverter-Based Distributed Generators. *IEEE Trans. Power Electron.* **2015**, *31*, 3600–3611. [CrossRef]
33. Xu, H.; Zhang, X.; Liu, F.; Shi, R.; Yu, C.; Cao, R. A Reactive Power Sharing Strategy of VSG Based on Virtual Capacitor Algorithm. *IEEE Trans. Ind. Electron.* **2017**, *64*, 7520–7531. [CrossRef]
34. Liu, Z.; Liu, J.; Zhao, Y. A Unified Control Strategy for Three-Phase Inverter in Distributed Generation. *IEEE Trans. Power Electron.* **2014**, *29*, 1176–1191. [CrossRef]

energies

MDPI

Article

Integrated Photovoltaic Inverters Based on Unified Power Quality Conditioner with Voltage Compensation for Submarine Distribution System

Desmon Petrus Simatupang and Jaeho Choi *

School of Electrical Engineering, Chungbuk National University, Cheongju 28644, Korea;
desmonpetrus@ymail.com
* Correspondence: choi@cbnu.ac.kr; Tel.: +82-43-261-2425; Fax: +82-43-276-7217

Received: 20 September 2018; Accepted: 23 October 2018; Published: 26 October 2018

Abstract: This paper proposes a multi-functional Photovoltaic (PV) inverter based on the Unified Power Quality Conditioner (UPQC) configuration. Power quality improvement is a difficult issue to solve for isolated areas or islands connected to the mainland through long submarine cables. In the proposed system, the line voltage is compensated for by the series inverter while the shunt inverter delivers the PV generating power to the grid. Depending on the technical conditions of power quality and system environment, it has five different operating modes. Especially during poor power quality conditions, the sensitive load is separated from the normal load to provide a different power quality level by using the microgrid conception. In this paper, the control method and the power flow for each mode are described, and the operational performance is verified through a PSiM simulation so that it can be applied to the power quality improvement of weak grid power systems such as in isolated areas or on islands connected to the mainland by long submarine cables.

Keywords: PV generation; submarine cables; shunt inverter; series inverter; voltage compensation; power quality

---

1. Introduction

Recently, renewable energy has spread rapidly due to concerns over climate change and environmental issues. The penetration of renewable energy in distribution systems has grown fast to catch up with population growth and reduce the use of conventional energy sources like coal power plants. Distributed generation (DG), such as photovoltaic (PV) systems or windtrubine generation systems, plays a major role in evolving a microgrid. PV energy has been developing for more than 160 years and its growth has increased exponentially in the last two decades. The total installed capacity at the end of 2016 amounted to 303 GW globally [1].

Since the PV system is designed to have significant penetration to the distribution line or submarine cable, the power flow can be reserved and the distribution system is no longer a passive circuit but an active one that can determine the power flow based on the distributed generation as well as the load. The low energy density and long-distance distribution system can make the grid system weak or its short circuit ratio (SCR) low [2,3].

People who live in isolated areas need electricity to manage their daily life. An islands' power system is usually connected to the mainland through a submarine cable system. However, the undervoltage or overvoltage problems can occur frequently due to the load capacity [4]. As in IEEE 1159, undervoltage is defined as a typical voltage magnitude less than 0.9 pu for a duration longer than 1 min, and overvoltage is defined as a typical voltage magnitude higher than 1.1 pu for a duration longer than 1 min [5]. In normal or heavy load conditions, undervoltage will occur due to cable resistance or cable inductance. On the other hand, in light loads, overvoltage can occur due to submarine cable capacitance—the so-called Ferranti effect [6,7]. In this condition, sensitive electric equipment can be damaged due to undervoltage or overvoltage grid conditions. In real measurements undertaken at Pramuka Island in Indonesia, the undervoltage was so serious that the line voltage decreased by up to 20% in the evening, but the overvoltage caused by the base load was not serious.

At the load side, some loads or equipment are sensitive to power quality problems. These sensitive loads can be differentiated from normal loads by a switch, so the microgrid technology allows each load group to be supplied power at different power quality levels [8]. To supply the electricity to sensitive loads with high power quality, additional power quality solutions (PQS), such as the active power filter or dynamic Uninterruptible Power Supply (UPS), can be installed for sensitive loads [9–11]. It is easy to apply all kinds of existing PQS at any location of grid to improve the power quality, but it requires a space for PQS and is a little bit more costly. As an alternative solution, it is possible to add a function of improving power quality to the power inverter of a microgrid [12]. The power inverter in the microgrid system is generally used for the conversion of the DC generating power to AC, so the main function of the inverter is to supply the converted power to the main grid under the output current control. However, the control of the inverter or the topology of the inverter itself can be modified to cover the additional function for the compensation of harmonics or voltage.

In recent years, the development of voltage compensation has led to the implementation of a Dynamic Voltage Restorer (DVR) [13–15]. The concept of using DVR for voltage mitigation has had positive results and has achieved popularity since its first use. The DVR with rechargeable energy was suggested to meet the power requirements for voltage disturbance mitigation [13,14], but the authors do not consider the penetration of renewable energy as a source of DVR. Several DVR strategies based on control strategies have also been developed in [15], but the authors do not consider the maximum voltage, which can be compensated for by DVR and the type of the load based on the voltage sensitivity.

The proposed PV inverter system in this research has the voltage compensation function, while the PV power is delivered to the grid. The configuration of the inverter is similar to that of the Unified Power Quality Conditioner (UPQC) [16,17]. It has the topology of the back-to-back inverter, in which one output is connected to the grid in a shunt and the other is connected to the grid in series. To improve the limited function of the conventional DVR in [13,14], the integrated PV system can supply energy to both the DVR and the grid. Therefore, instead of compensating for the harmonics or reactive power in the UPQC, the generated PV power is delivered to the grid through the shunt inverter while the series inverter compensates for the voltage when undervoltage or overvoltage is occurred in the grid [18,19].

To improve the power quality at the system level, all loads are divided into normal loads and sensitive loads groups by integrating the microgrid concept to maintain the power quality at different levels [20]. The voltage compensation operation should be performed only for the sensitive loads group. Depending on the state of the grid voltage, the voltage is compensated for under the grid-connected mode or the regulated voltage is generated under the stand-alone mode without connection to the grid [21]. Therefore, if the grid voltage is in the normal range, the shunt inverter operates only in the PV generation mode. If the grid voltage is outside of the normal range, the grid voltage for only sensitive load is compensated for by the series inverter while the generated PV power is delivered to the grid through the shunt inverter. On the other hand, if the grid voltage deviates from the voltage compensation range, the sensitive loads group is disconnected from the grid and is supplied the power from the PV systems through the shunt inverter. In this mode, it is not necessary to compensate for the voltage through the series inverter, and the normal loads group received power from the grid continuously although the power quality was very poor.

The topology and operation of the PV inverter, which has the additional function of voltage compensation, depending on the state of grid voltage, are described in this paper. The operation mode is defined as a normal mode, a voltage compensation mode, and the stand-alone mode when the grid voltage is in the range of ±10%, ±10 to ±20%, or over ±20%, respectively. During the low irradiance condition, the Battery Energy Storage System (BESS) [22,23] can be a power source in place of the PV source. However, the operation of BESS is applied only for the voltage compensation mode or the stand-alone mode considering the capacity of the batteries. In the voltage compensation mode, the voltage compensation is performed through the series inverter. On the other hand, in stand-alone mode, the shunt inverter supplies the power to the grid without voltage compensation through the series inverter. To analyze the performance of the proposed voltage compensation method, a simulation based on PSIM is conducted.

In this paper, the modeling and problems caused by the submarine cable are described in Section 2. The topology of the proposed PV system with series and shunt inverter is shown in Section 3. In Section 4, the control algorithm and the power flow in every case are explained. Finally, the simulation results and analyses are reported in Section 5, with conclusions in Section 6.

2. Modeling and Analysis of Submarine Cable

2.1. Modeling of Submarine Cable

A submarine cable is installed under the sea to supply power to an isolated island from the mainland or connect an offshore wind turbine system to the main grid. Since it is located near the ground level and has thick insulation, a submarine cable has better capacitive characteristics than an overhead line. Based on a theoretical parallel plate capacitor, the capacitance is inversely proportional to the gap between plates, which corresponds to the distance from the earth in this case. Other factors, such as the distance to other cables (relatively short compared to the overhead line) and the capacitance between conductor and insulation cable, make the submarine cable have better capacitive characteristics than an overhead line.

As with the transmission line, the submarine cable can be represented sufficiently well by the resistance (R), inductance (L), and capacitance (C). The difference between the transmission line and the submarine cable is in the short length. The transmission line assumed the shunt admittance of the line is small and can be neglected, but in the submarine cable, the shunt admittance is relatively high and can have an impact on the calculation, as shown in Figure 1. Therefore, in a short submarine cable, the shunt admittance is included in the calculation.

For short and medium lengths, lumped parameters are used, which give better accuracy. However, for the exact solution of any lines longer than 240 km, it should be considered that the parameters of the line are not lumped but distributed uniformly in the line [24].

Cable Impedance

$$R + jX$$

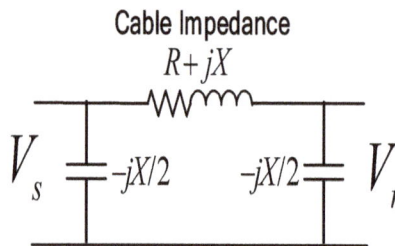

Figure 1. Modeling of short and medium-length submarine cables.

2.2. Voltage Profile of Submarine Cable

Most of the submarine systems for loads on isolated islands are radial configuration, which gives high impedances. These high impedances lead to the voltage drop or Ferranti effect along the cable from the sending end to the receiving end load. The voltage drop can be calculated from the basic analysis of a two-bus distribution system, as in Figures 2 and 3.

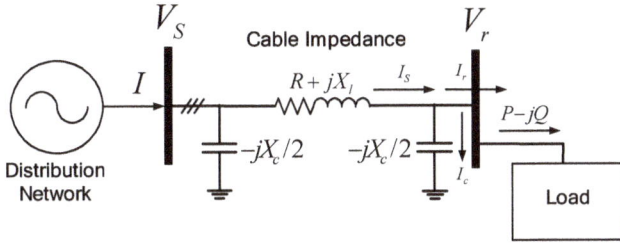

Figure 2. Distribution system submarine cable modeling without distributed generation (DG).

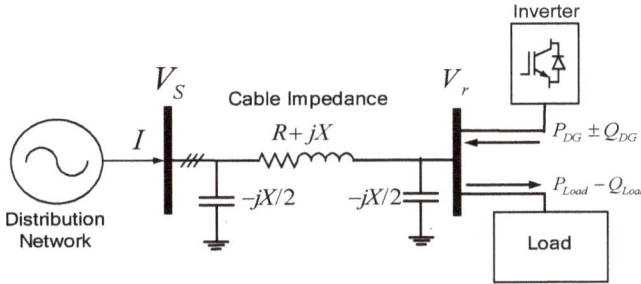

Figure 3. Distribution system submarine cable modeling with DG.

The relationship between the sending end bus voltage, V_s, and the receiving end bus voltage, V_r, is given as in Equation (1):

$$V_s = I_s(R + jX_l) + V_r = (I_r + I_c)(R + jX_l) + V_r. \tag{1}$$

Substituting I_r and I_c, described as in Equation (2), into Equation (1), we get Equation (3):

$$I_r = \frac{P_{load} - jQ_{load}}{V_r} \quad I_c = \frac{V_r}{-2jX_c} \tag{2}$$

$$V_s = \left(\frac{RP_{load} + X_lQ_{load}}{V_r} - \frac{V_rX_l}{2X_c} + V_r\right) + j\left(\frac{X_lP_{load} - RQ_{load}}{V_r} + \frac{RV_r}{2X_c}\right). \tag{3}$$

If the phase difference between V_s and V_r is neglected based on V_s, then the voltage difference between V_s and V_r can be described as in Equation (4):

$$\Delta V_r = V_s - V_r = \frac{RP_{load} + X_lQ_{load}}{V_r} - \frac{V_rX_l}{2X_c}. \tag{4}$$

This shows that the bus voltage can increase or decrease depending on the amount of active and reactive power consumed at the load and supplied from the Distributed Generator (DG). Based on Equation (4), two conditions can result depending on the load capacity. The voltage difference in Equation (4) can be categorized into two parts: The first part is $\frac{RP_{load} + X_lQ_{load}}{V_r}$, and the second part is $-\frac{V_rX_l}{2X_c}$. In the heavy load condition, the first part is dominant, and the voltage difference, ΔV_r, is positive, which means that there is a voltage drop at the receiving end. On the other hand, in light or no-load conditions, the second part is dominant, and the voltage difference, ΔV_r, is negative, which means that there is a voltage rise at the receiving end [25]. This is the so-called "Ferranti Effect".

The PV penetration in the distribution system can increase the voltage in the Point of Common Coupling (PCC) bus depending on the power injected by the PV system. Figure 3 shows how the DG can change the voltage at the receiving end since the DG and the power electronics technology can control the active and reactive power injection. In this case, Equation (4) can be modified as in Equation (5):

$$\Delta V_r = V_s - V_r = \frac{R(P_{Load} - P_{DG}) + X(Q_{Load} - Q_{DG})}{V_r} - \frac{V_r \omega^2 LC}{2}. \tag{5}$$

3. Configuration and Operation of Proposed PV Inverter with Voltage Compensation

3.1. Proposed PV Inverter

The conventional PV system delivers the power to the grid directly in the grid-connected mode or supplies the power to the load in stand-alone mode without any ability to improve the power quality. However, the proposed system, which has the back-to-back inverter topology of shunt and series inverters based on UPQC configuration, has the additional function of voltage compensation to the conventional PV system, as shown in Figure 4. The sensitive load can be separated from the normal load by disconnecting the switch, and it can be supported at a different power quality level from the normal load by the voltage compensation of the series inverter.

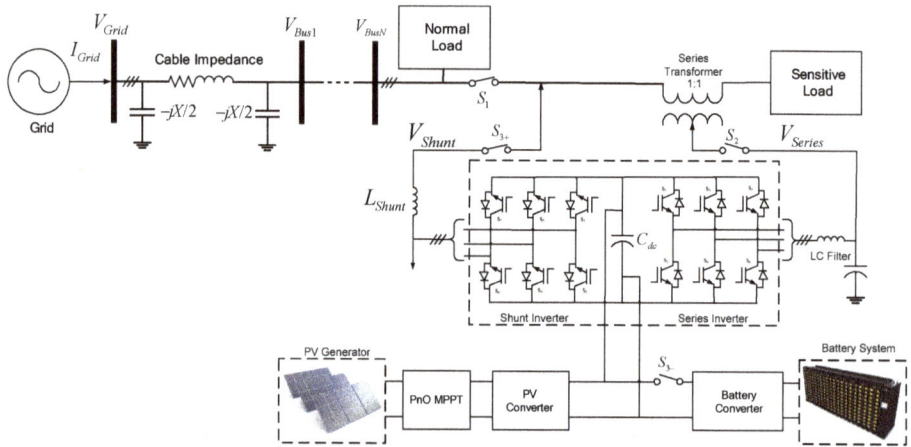

Figure 4. Proposed PV system for voltage compensation method.

3.2. Operation of Series Inverter

The series inverter is used to compensate for the voltage in the sensitive load in a weak grid system [26–30]. During the light load condition, as shown in Figure 5a, overvoltage, V'_G, is occurred, and then the series inverter injected the negative voltage, $V'_{series\ inverter}$, into the grid to mitigate it. On the other hand, during heavy load conditions, as shown in Figure 5b, undervoltage, V'_G, occurs, so the series inverter injects the positive voltage, $V'_{series\ inverter}$, into the grid to mitigate it.

After applying the in-phase voltage compensation method, which makes the series inverter voltage synchronized to the grid voltage, an exchange of active and reactive power between the series inverter and the grid occurs [31]. The voltage and power calculation for series inverter in-phase control scheme is as follows:

$$V'_{series} = \sqrt{2}\left[V_{load} - V'_{grid,abc}\right] \tag{6}$$

$$P_{series} = \frac{3}{2}\left[V_{d,series}I_{d,series} + V_{q,series}I_{q,series}\right] \tag{7}$$

$$Q_{series} = \frac{3}{2}\left[V_{d,series}I_{q,series} - V_{q,series}I_{d,series}\right]. \tag{8}$$

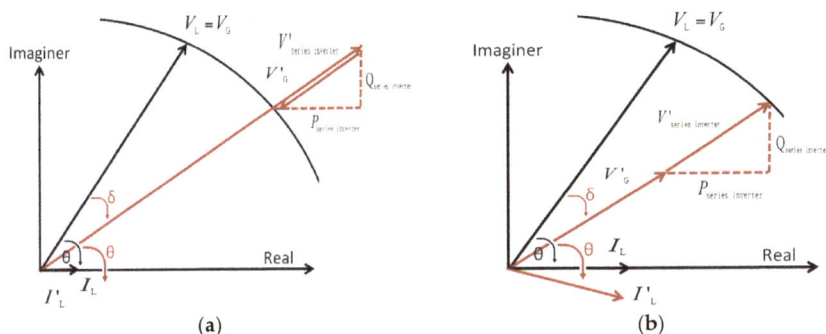

Figure 5. Phasor diagram for the series inverter: (**a**) overvoltage conditions; (**b**) undervoltage conditions.

3.3. Operation of Shunt Inverter

In this proposed method, the operation of the shunt inverter can be divided into four functions. Firstly, in normal voltage conditions, the shunt inverter injects the maximum power generated from the PV by the Maximum Power Point Tracking (MPPT) method to the grid. Secondly, in undervoltage or overvoltage conditions, the maximum power injected to the grid depends on the power exchanged through the series inverter. In undervoltage conditions, the maximum power injected by the shunt inverter is reduced due to the power necessary for the voltage compensation through the series inverter. On the other hand, in overvoltage conditions, the series inverter absorbs the power from the grid, so the maximum power injected to the grid through the shunt inverter is increased due to the power absorbed from the grid through the series inverter. Thirdly, if the grid voltage is outside the range of voltage compensation or the battery back operation is necessary due the low irradiance, the sensitive load is isolated from the grid by disconnecting the switch. Then, the shunt inverter fully covers the sensitive load in stand-alone mode while the grid supplies power to the normal load even if the power quality is poor. Finally, if the State of Charge (SoC) of BESS is lower than the lower limit during low irradiance conditions, the shunt inverter absorbs power from the grid side and delivers it to the series inverter for voltage compensation. In this mode, the load will be increased due to the power received through the shunt inverter, and this load increase may make the voltage drop in the grid worse. The delivered power through the shunt inverter is described as follows:

$$P_{shunt} = \frac{3}{2}\left[V_{d,shunt}I_{d,shunt} + V_{q,shunt}I_{q,shunt}\right] \tag{9}$$

$$Q_{shunt} = \frac{3}{2}\left[V_{d,shunt}I_{q,shunt} - V_{q,shunt}I_{d,shunt}\right]. \tag{10}$$

4. System Operation Mode and Control of Proposed PV Inverter

There are five operation modes in the proposed PV inverter system, depending on the irradiance condition and the grid voltage magnitude: the normal mode, the voltage compensation mode with PV power, the stand-alone mode, the voltage compensation mode with battery energy, and the voltage compensation mode with the grid power.

4.1. Normal Mode

When the grid voltage is under the ±10% range of the rated value, the PV inverter is in the normal mode. In the normal mode, the shunt inverter delivers power to the grid based on the MPPT scheme without any voltage compensation, but the series inverter does not work in this mode. The control scheme for the shunt inverter is shown in Figure 6a. The maximum active power is set to be a reference value of the active power for the PQ control. On the other hand, the reference reactive power is set to be zero to meet the unity power factor operation. The reference values for the active and reactive powers are converted to the reference values of I_d and I_q by Equations (11) and (12). Hereinafter, the reference values, I^*_d and I^*_q, are subtracted from the real values, I_d and I_q, and the PI control is applied to regulate I_d and I_q to follow the reference values, I^*_d and I^*_q. Finally, the reference dq-frame signal is be transformed to abc domain for Pulse Width Modulation (PWM) generation. The power flow during this mode is shown in Figure 6b.

$$I*_{d,PQ} = \frac{\frac{2}{3}P*_{PQ} + V_{q,PQ} \cdot I_{q,PQ}}{V_{d,PQ}} \tag{11}$$

$$I*_{q,PQ} = \frac{\frac{2}{3}Q*_{PQ} + V_{q,PQ} \cdot I_{d,PQ}}{V_{d,PQ}} \tag{12}$$

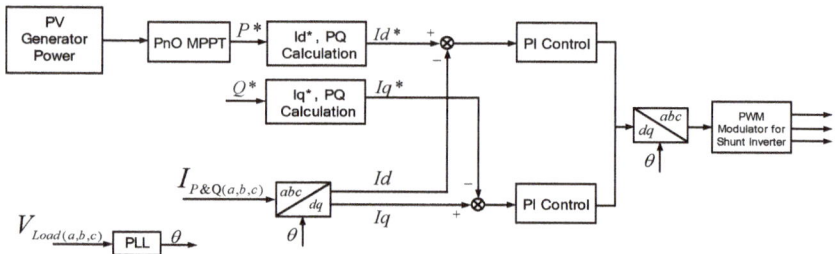

(a)

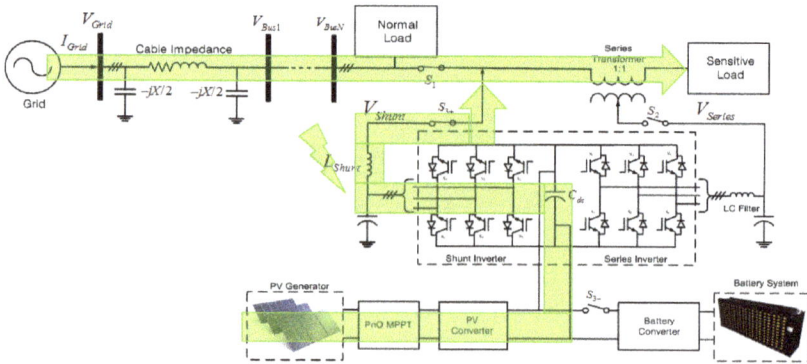

(b)

Figure 6. Normal mode: (a) Control scheme for shunt inverter; (b) power flow condition.

4.2. Voltage Compensation Mode with PV Power

When the grid voltage is between ±10% and ±20% of the rated voltage, it is in the voltage compensation mode. Then the line voltage is compensated for by the series inverter while the PV power is mostly delivered by the shunt inverter. In the voltage compensation mode, the series inverter injects the voltage into the grid which is in phase with the grid voltage after fault. In this case, the series inverter injects the minimum magnitude voltage compared to other series inverter methods like pre-sag compensation and quadrature injection [31].

Figure 7a,b shows the control blocks for voltage compensation. Figure 7a shows the control scheme for shunt inverter. The reference currents, I^*_d and I^*_q, are generated to deliver the power through the shunt inverter, the same as the power difference between the PV generated power and the power delivered to the grid through the series inverter. Then the current-controlled PWM is applied to let the output current of shunt inverter, I_d and I_q, follow this reference value. On the other hand, the reference voltages, V^*_d and V^*_q, are generated to supply the compensation voltage, and the voltage-controlled PWM is applied to regulate the output voltage of the series inverter, V_d and V_q, to follow this reference value, as shown in Figure 7b. The power flow during this mode is shown in Figure 7c.

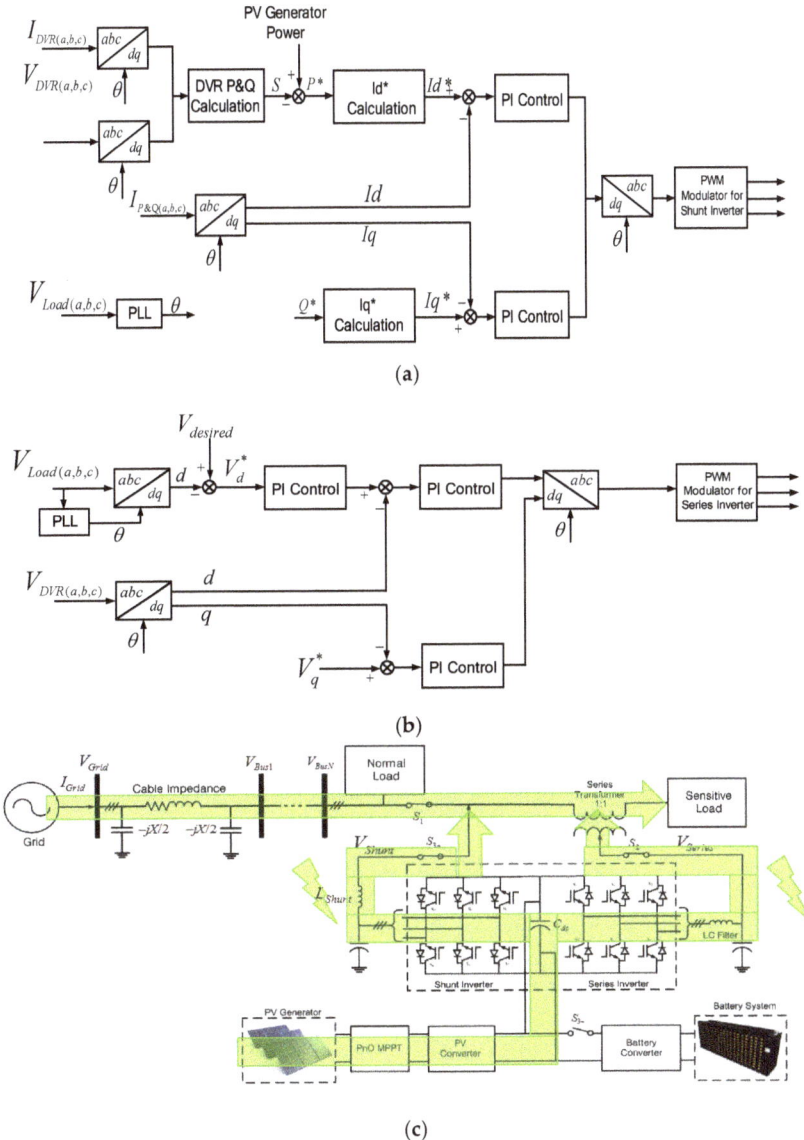

Figure 7. Voltage compensation mode: (**a**) Control scheme for shunt inverter; (**b**) control scheme for series inverter; (**c**) power flow condition.

4.3. Stand-Alone Mode

When the line voltage is out of range for compensation, the static switch disconnects the sensitive load from the grid, and then the PV power is delivered to the load by the shunt inverter under the stand-alone mode without any voltage compensation by the series inverter, as shown in Figure 8a,b.

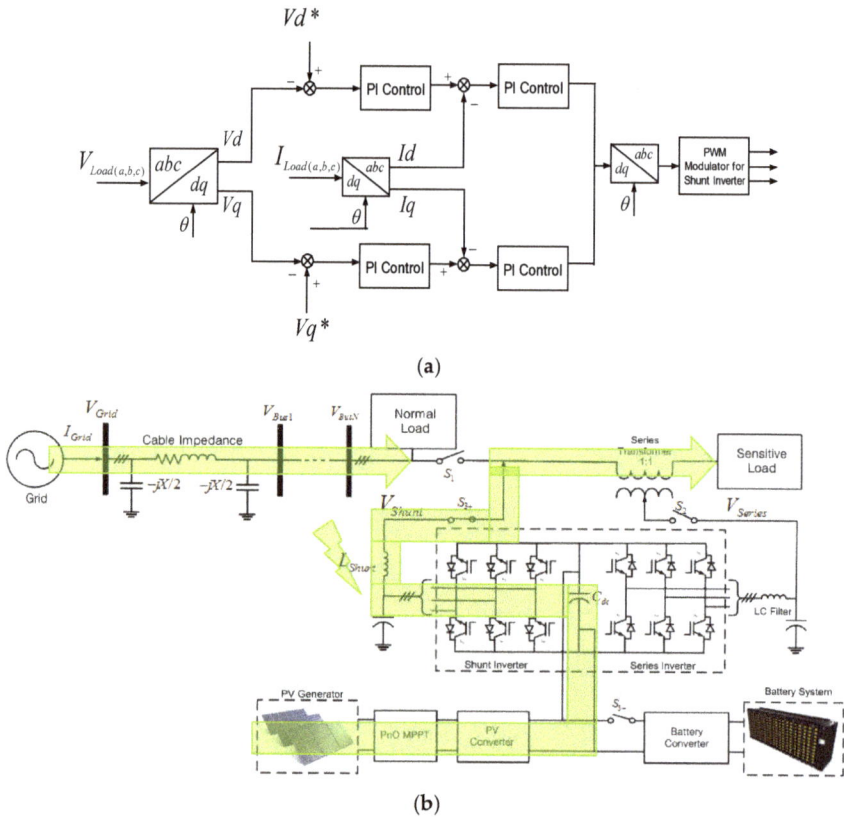

(a)

(b)

Figure 8. Stand-alone mode: (a) Control scheme; (b) power flow condition.

The load voltage, V_d and V_q, is controlled to follow the reference voltage, V^*_d and V^*_q. This controller is composed of the outer voltage controller and the inner current controller. Then, the output abc signals are fed to PWM to generate the gating pulses for the shunt inverter.

For transferring between the power control under the grid connected mode and the voltage control under the stand-alone mode, the multiplexer is installed before the dq to abc transformation. The multiplexer may multiply the signal in voltage control to be zero when the grid voltage is in the range of the grid connected. The same treatment can be applied in power control, so the multiplexer will make the power control signal have a zero value when the grid voltage is in the range of the stand-alone mode.

4.4. Voltage Compensation Mode from BESS

In low irradiance or at night, the PV cannot produce power, so the BESS system provides the power for the series inverter only as shown in Figure 9a,b. In undervoltage conditions, the BESS will not supply the power to the shunt inverter due to the limitations of BESS capacity. On the other hand, the BESS system supports the series inverter only to mitigate the voltage disturbance in this mode.

The control scheme is the same as the voltage compensation mode of the series inverter; the only difference is the power source. In the voltage compensation mode with PV power, the power source is the PV power, but in this mode, the power source is the BESS power. The BESS capacity SoC in this simulation is assumed to be 1 or fully charged.

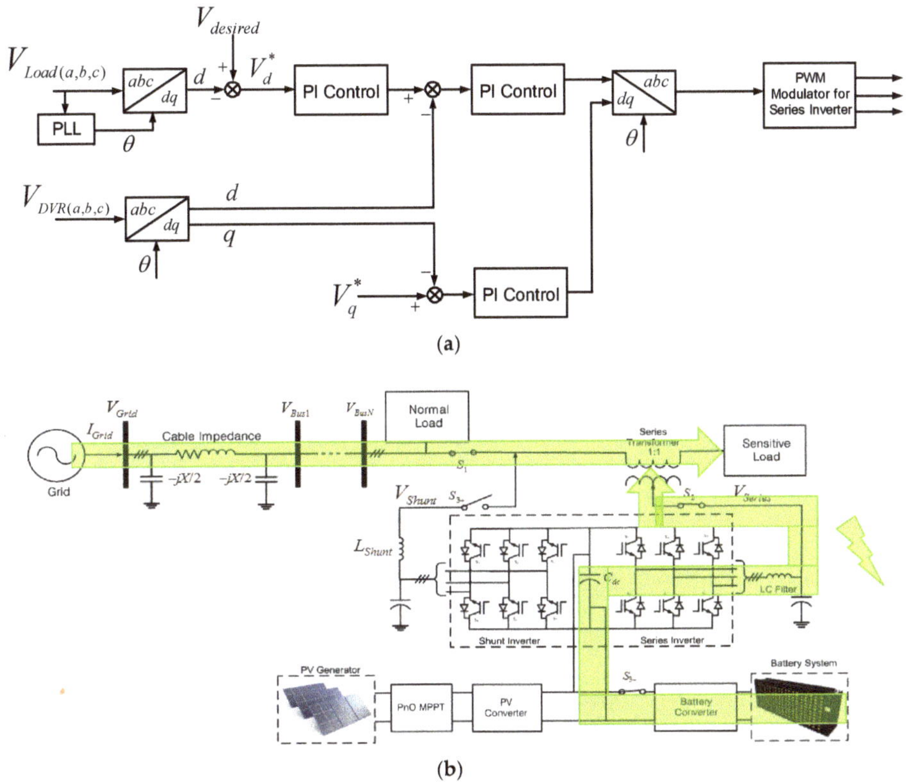

(a)

(b)

Figure 9. Voltage compensation mode from BESS: (**a**) Control scheme; (**b**) power flow condition.

4.5. Voltage Compensation Mode from Grid

In this condition, we assumed that the BESS power almost reached 0.1 SoC. There will be times when the BESS cannot support the series inverter anymore. In this condition, the BESS switch will be turned off. The shunt inverter will be a load from the grid side to absorb the power and deliver it to the series inverter, as shown in Figure 10. Thus, the series inverter can compensate for the undervoltage in this condition. The control scheme for the series inverter is the same as the voltage compensation mode. For the shunt inverter, there are some modifications due to supply power to the series inverter. The power absorbed from this mode is based on the power required by the series inverter. Thus, the value of the power desired in this condition is based on Equations (11) and (12), which will be determined by the demand from the series inverter.

(a)

(b)

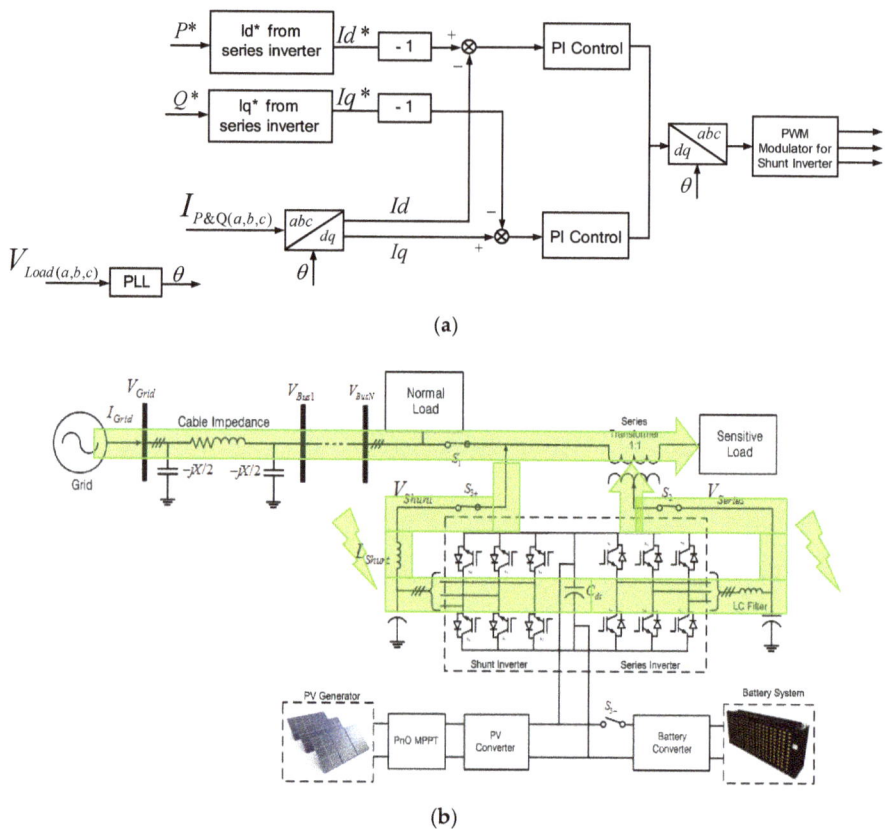

Figure 10. Voltage compensation mode from grid: (**a**) Control scheme; (**b**) power flow condition.

4.6. Automatic Transfer Switches

There are four switches in this proposed scheme. Switch 1, S_1, is located between the grid and the sensitive load; Switch 2, S_2, is between the series inverter and the sensitive load; Switch 3, S_{3+}, is between the shunt inverter and the grid; and Switch 4, S_{3-}, is between the BESS and inverters, respectively. The control scheme of switches for series and shunt inverters are based on the value of the grid voltage. If the sensitive load voltage is 0.9 to 1 pu, S_1 is turned on. The shunt inverter is connected to the grid and supplies the power to the sensitive load without voltage compensation, as shown in Figure 11a,c. If the voltage magnitude is 0.8 to 0.9 pu, Switch S_2 is turned on and the voltage compensation is applied by the series inverter, while the shunt inverter delivers the generated PV power to the grid through the shunt inverter. If the line voltage is lower than 0.8pu rated voltage, then both S_1 and S_2 are turned off, as shown in Figure 11a, to separate the sensitive load from the normal load. Then it operates as the stand-alone mode and the shunt inverter supplies the power directly to the sensitive load under the voltage control, while the normal load is supplied power from the grid with poor power quality.

The same control strategy can be applied for overvoltage condition. If the line voltage is between 1.1 and 1.2 pu, Switches S_1 and S_2 are turned on and the series inverter compensates for the overvoltage while the shunt inverter delivers the PV power to the grid. When the line voltage is higher than 1.2 pu, Switches S_1 and S_2 are turned to the off position and the stand-alone mode is applied, the same as in undervoltage conditions.

On the other hand, during a no PV generation condition, the BESS supports the series inverter without the operation of the shunt inverter by turning off Switch S_{3-}. However, when the SoC of BESS is lower than 0.1, the shunt inverter absorbs the power from the grid to support the series inverter by turning on Switch S_{3+}.

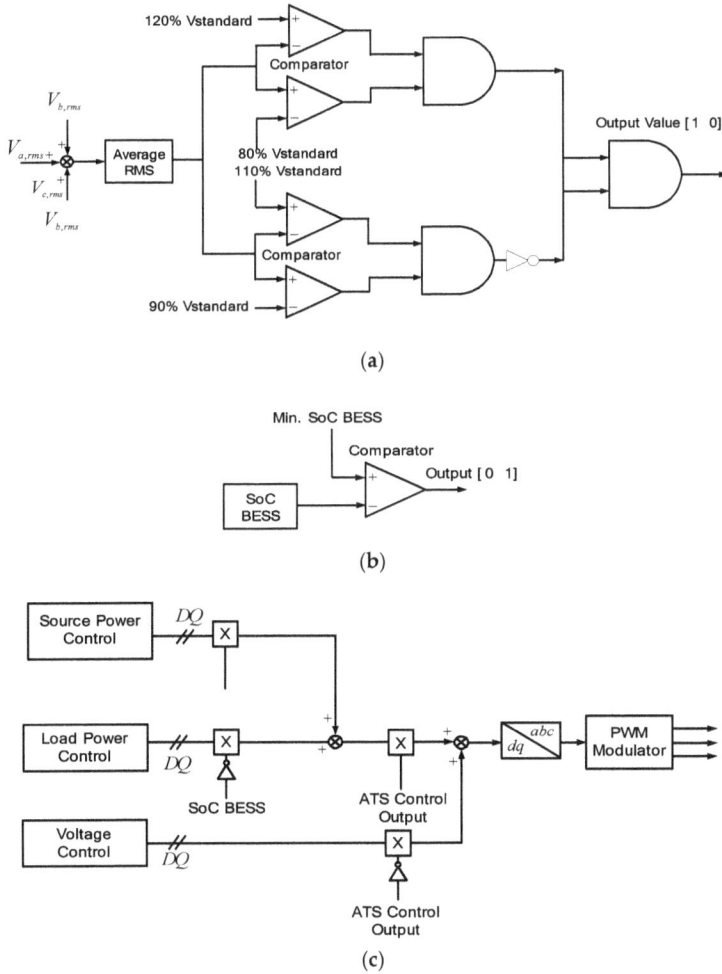

Figure 11. Automatic transfer switches: (**a**) Series inverter; (**b**) BESS; (**c**) shunt inverter.

5. Simulation Results and Discussion

5.1. Simulation Conditions and Parameters

In this simulation, the PV with MPPT model, the series inverter, the shunt inverter, the grid, and the loads are presented. Table 1 and Figure 12 present the simulation parameters and the system operation flowchart for PSIM simulation. The PV model is implemented by using the physical model, which is available in PSiM software (version, city country). The BESS is lithium-ion and the model is available in PSiM software.

Each case has been tested in several conditions. For the undervoltage case, the normal condition is shown first, followed by the grid undervoltage between 0.8 to 0.9 pu and the stand-alone mode, which is the grid undervoltage less than 0.8pu. For the overvoltage, the normal condition is presented from t < 1 s, followed by the grid overvoltage in the range of 1 to 1.1 pu and the stand-alone mode when the grid voltage is more than 1.1 pu.

The next case is the voltage compensation without PV power mode. In this condition, it is assumed that the BESS can supply the power to the series inverter, so the BESS generates the power to the series inverter only for voltage compensation. The last case is when the BESS cannot support the series inverter anymore, so the shunt inverter delivers the power from the grid to the series inverter.

Table 1. Simulation parameters.

Part of Circuit	Descriptions	Value
Grid	Main Voltage	220; 190; 150 V 220; 230; 264 V
	Frequency	50 Hz
Battery	Capacity	50 Ah
Undervoltage Load	Sensitive Load Active	80 kW
	Sensitive Load Reactive	60 kVar
Overvoltage Load	Sensitive Load Active	72 kW
	Sensitive Load Reactive	50 kVar
DC Capacitor	Capacity	$C_{dc} = 10$ mF
PV System	Max Power	113 kW
	Open Circuit Voltage	684 V
	Short Circuit Current	180 A
	V_{mpp}	675 V
	I_{mmp}	168 A
Series Connection	Transformer	1:1
	Filter	$C_{se} = 50$ µF, $L_{se} = 1$ mH
	PI-1, PI-2 & PI-3	20
Shunt Connection	Filter	$C_{sh} = 20$ µF, $L_{sh} = 2$ mH
	PI-1 & PI-2 (Power Control)	200
	PI-1 & PI-2 (Voltage and Load Mode)	10
PWM	Sampling Frequency	5 kHz

5.2. Undervoltage Case

In this simulation, there are three cases within certain periods, as shown in Figure 13. Before 1 s, no undervoltage occurred. Therefore, there is no action from the series inverter while the shunt inverter injects the active power to the grid. During 1 < t < 2 s, the undervoltage occurs under 0.9 pu. The switch S_2 is turned on and the series inverter injects the voltage to compensate for the grid voltage. After t > 2 s, if the voltage is lower than 0.8 pu, the shunt inverter takes care of the sensitive load in stand-alone mode while the grid covers the normal load in grid mode. In Figure 13c, it is verified that the voltage waveform of the sensitive load is well regulated during the undervoltage condition by the voltage compensation of the series inverter. In this period, the shunt inverter control is transferred from the power control mode to the voltage control mode.

The simulation results for power flow are shown in Figure 14. During the period t < 1 s, the grid injects power for the normal condition as shown in Figure 14a. In Figure 14b,c, there is no power flow through the series inverter, which is in the off position, but the shunt inverter injects the active power generated by PV to the grid system. When undervoltage occurs during 1 < t < 2 s, the grid power is decreased due to the voltage drop, as shown in Figure 14a. The series inverter injects the power to compensate for the undervoltage, while the remaining PV generating power flows through the shunt inverter, as shown in Figure 14b,c. The active and reactive power of sensitive load are constant in this period since the series inverter compensates for the undervoltage through the series inverter as shown in Figure 14d,e. During the last period (t > 2 s), the grid voltage is less than 0.8 pu. The series inverter is in off mode and the shunt inverter takes care of the sensitive load in the stand-alone mode. Therefore, the normal load is covered by the grid as shown in Figure 14a,d,e, while the shunt inverter supplies the power to the sensitive load, separate from the normal load, as shown in Figure 14c–e.

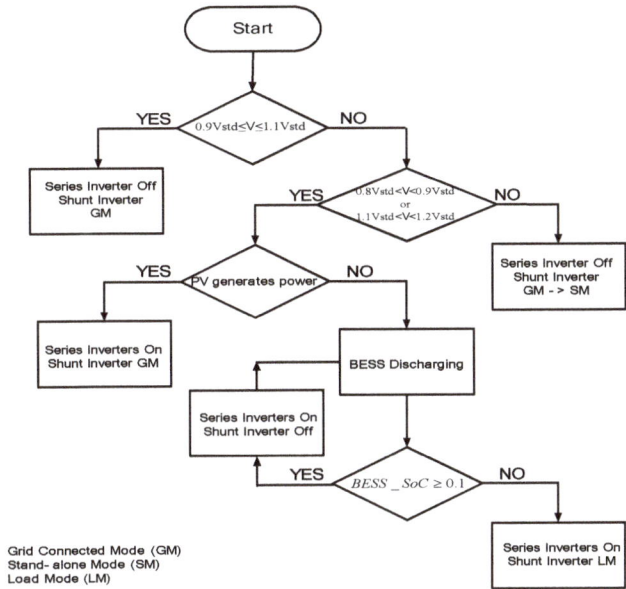

Figure 12. Flowchart for proposed PV system for voltage compensation method.

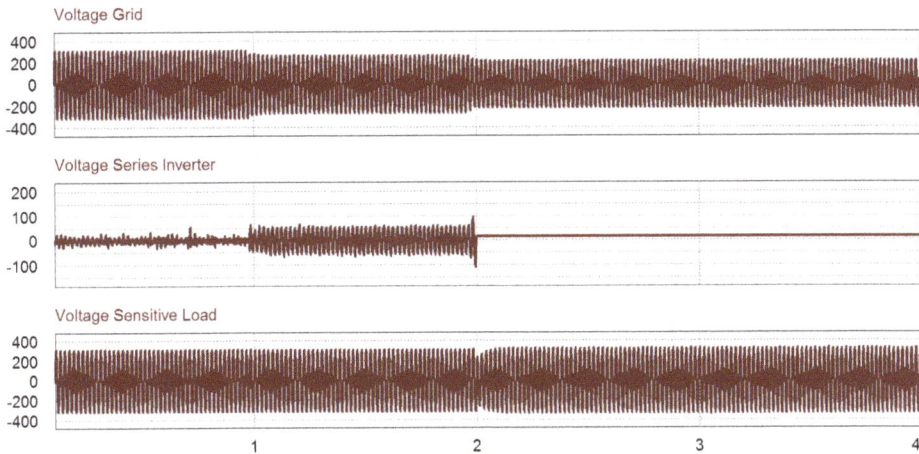

Figure 13. Simulation results of undervoltage case: (**top**) Grid voltage; (**middle**) series inverter voltage; (**bottom**) sensitive load voltage.

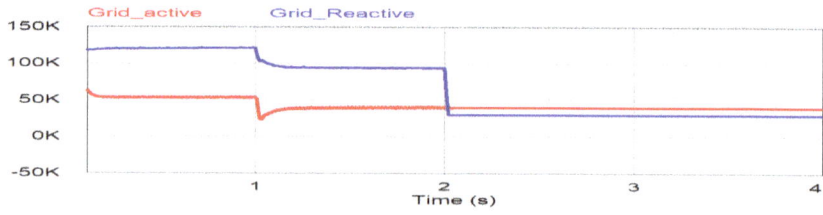

(a)

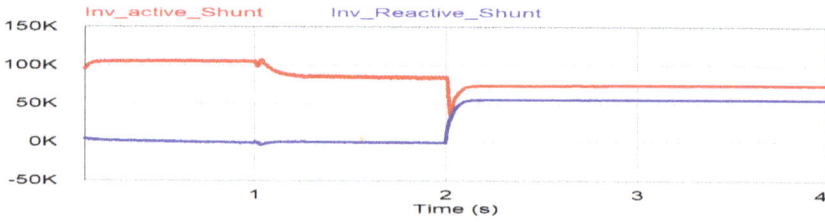

(b)

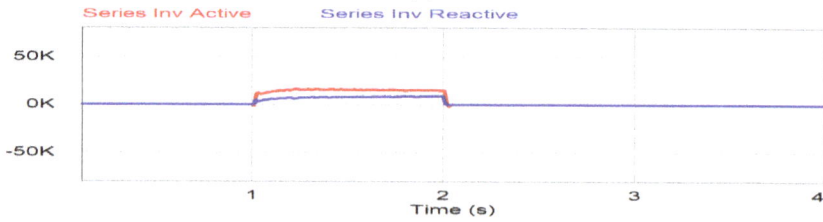

(c)

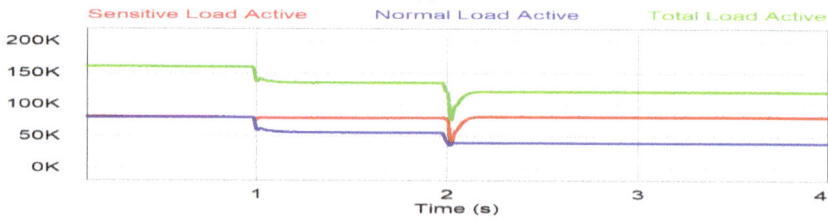

(d)

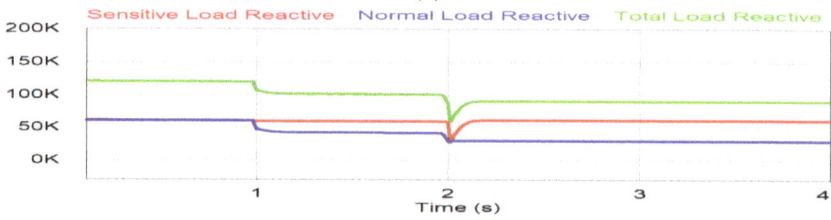

(e)

Figure 14. Simulation results of undervoltage: (**a**) Grid power; (**b**) shunt inverter power; (**c**) series inverter power; (**d**) active power for sensitive, normal and total load; (**e**) reactive power for sensitive, normal and total load.

5.3. Overvoltage Case

The performance analysis in the overvoltage case is applied in reverse to the under–voltage case. As shown in Figure 15, the series inverter does not operate under the normal before 1 s while the shunt inverter injects the active power to the grid. During 1 < t < 2 s, the overvoltage occurs over 1.1 pu, then the series inverter compensates for the voltage by absorbing the power from the grid. After t > 2 s, the voltage is higher than 1.2 pu, then the shunt inverter keeps the voltage for the sensitive load in stand-alone mode. In Figure 15c, it is verified that the voltage waveform of sensitive load is well regulated during the overvoltage condition by the voltage compensation of the series inverter.

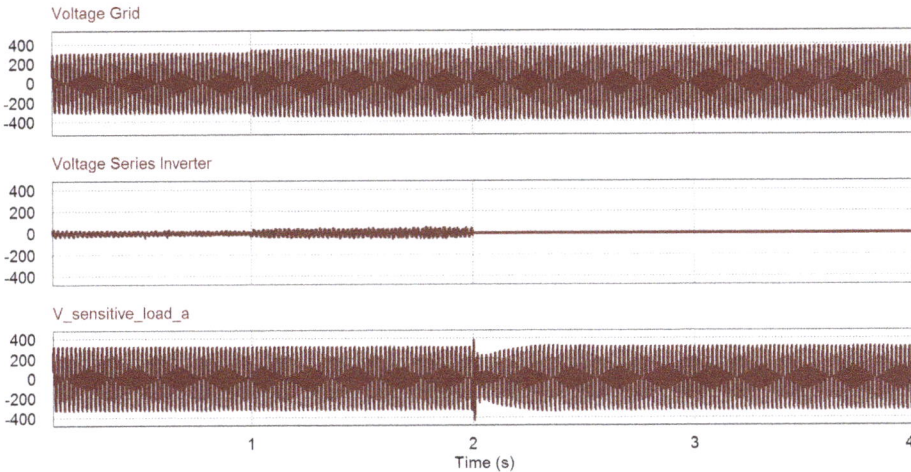

Figure 15. Simulation results of overvoltage case: (**top**) Grid voltage; (**middle**) series inverter voltage; (**bottom**) sensitive load voltage.

The power flow under the overvoltage case is same as that of undervoltage case as shown in Figure 16. Before 1 s, the grid is under the normal condition, so there is no power flow through the series inverter while the shunt inverter injects the generated PV power to the grid system. When the overvoltage occurs during 1 < t < 2 s, the series inverter compensates for the voltage while the remaining PV generating power flows through the shunt inverter as shown in Figure 16b,c. The active and reactive power of sensitive load is constant in this period since the series inverter compensates for the overvoltage through the series inverter as shown in Figure 16d,e. During the last period (t > 2 s), the grid voltage is higher than 1.2 pu. The series inverter is in off mode and the shunt inverter takes care of the sensitive load in the stand-alone mode. Therefore, the normal load is covered by the grid as shown in Figure 16a,d,e, while the shunt inverter supplies the power to the sensitive load only, as shown in Figure 16c-e.

5.4. Voltage Compensation without PV Power Case

This case is the condition of low irradiance or during the night. Figure 17a shows the voltage compensation after 1 s by discharging the battery. Before 0.5 s, the PV-generated power is supplied to the grid through the shunt inverter, so the grid injects less active power to the grid due to the power delivery of shunt inverter shown in Figure 17a-c. During 0.5 to 1.0 s, the PV cannot generate active power due to the low irradiance condition and the grid injects more power to the loads shown in Figure 17a. After t > 1.0 s, undervoltage occurs, so the BESS supplies power to the series inverter to compensate for the voltage shown in Figure 17b,c. The active and reactive power for the sensitive load can stay constant, as shown in Figure 17d. During this range, the BESS can be the power source of the series inverter instead of the PV power source, so the SoC of BESS is reduced due to the power supply to the series inverter, as shown in Figure 17e.

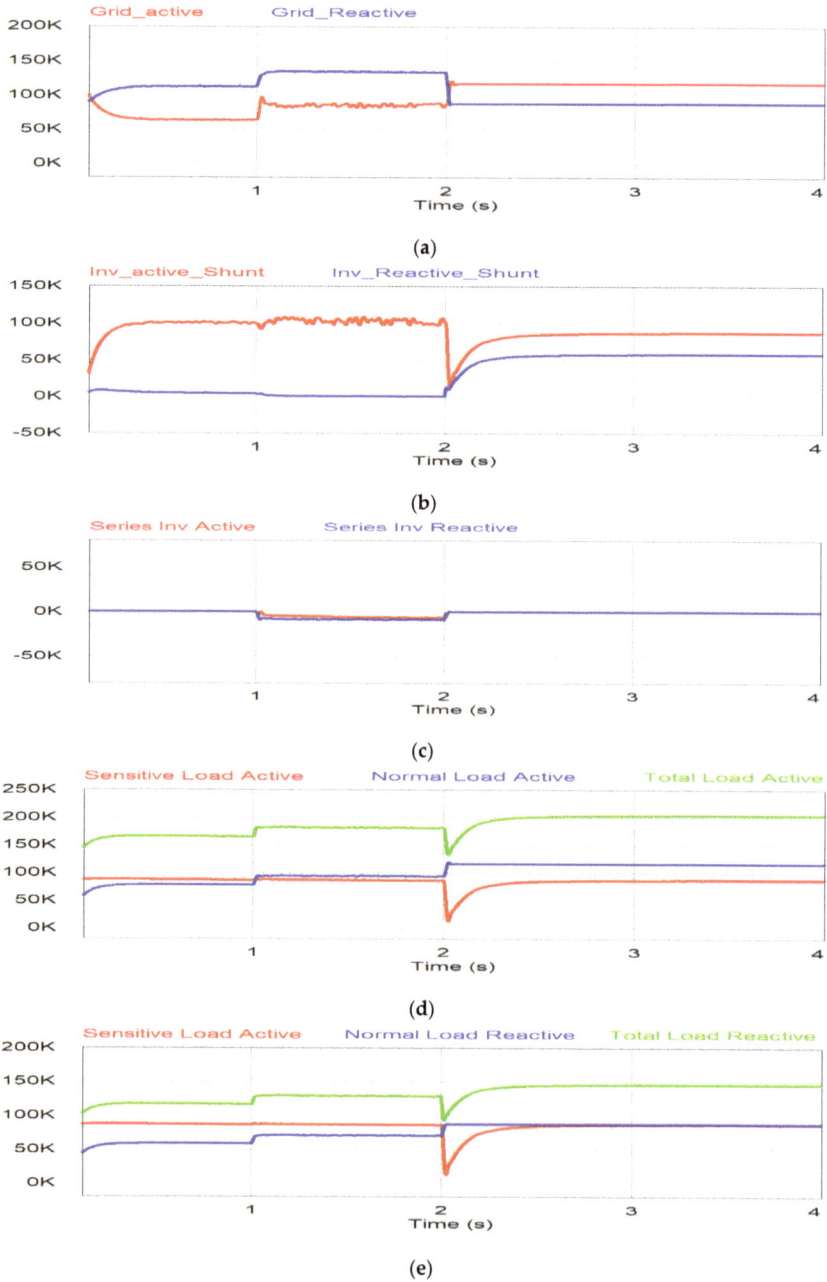

(a)

(b)

(c)

(d)

(e)

Figure 16. Simulation results of overvoltage: (**a**) Grid power; (**b**) shunt inverter power; (**c**) series inverter power; (**d**) active power for sensitive, normal and total load; (**e**) reactive power for sensitive, normal and total load.

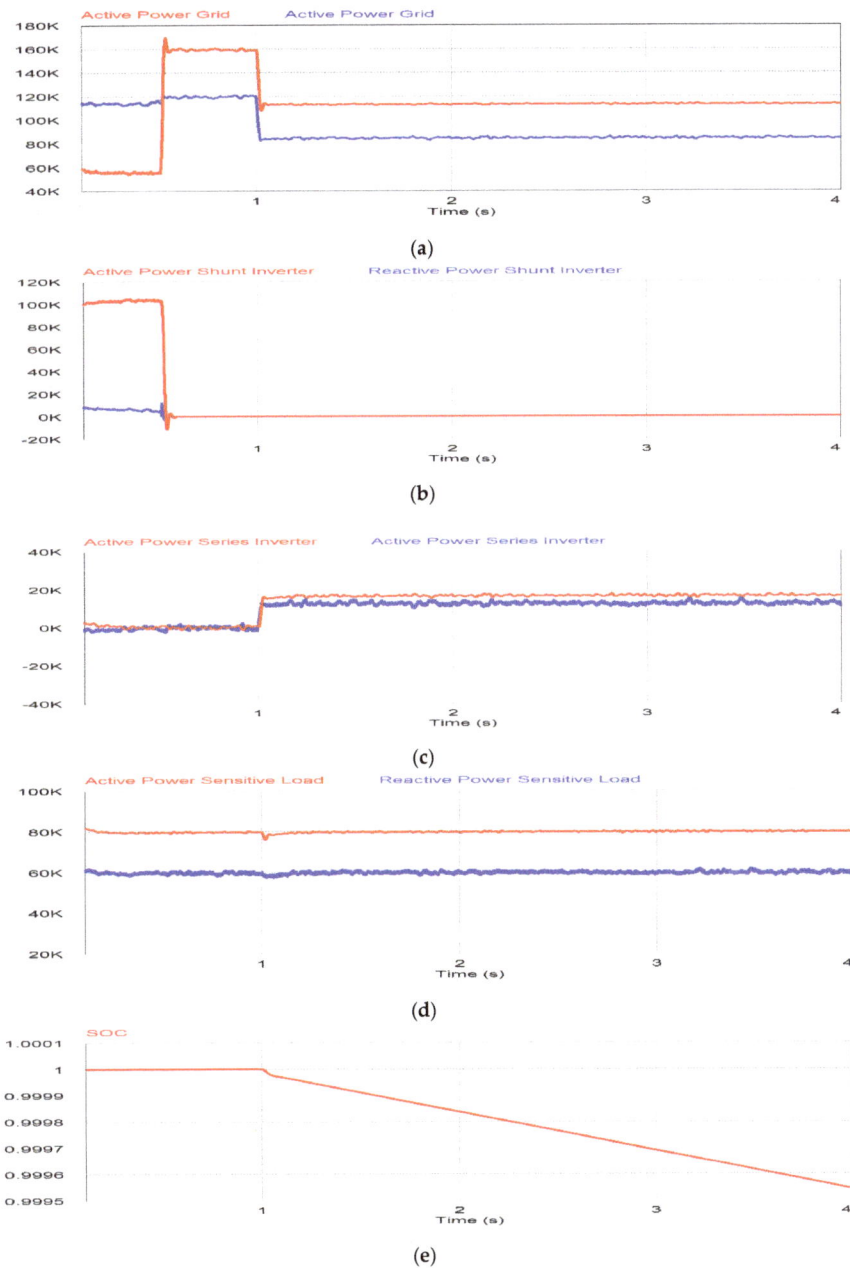

Figure 17. Simulation results of BESS: (**a**) Grid power; (**b**) shunt inverter power; (**c**) series inverter power; (**d**) active and reactive power for sensitive load; (**e**) SoC of BESS.

5.5. Voltage Compensation with Power from Grid Case

The last mode is the voltage compensation with the grid power mode as shown in Figure 18 and the continuation of the case after the voltage compensation without PV power case. Before t < 2 s, the BESS has to support the series inverter due to no PV power generation as shown in Figure 18b,e, and the response is the same of the voltage compensation without PV power after 1 s. Switch S_{3+} is in the off position and Switch S_{3-} is in the on position. The grid still supplies the sensitive load, though the grid voltage is in the undervoltage condition, as shown in Figure 18a,d. When the SoC of BESS reaches 0.1, as shown in Figure 18e, the BESS cannot supply the power to the series inverter anymore under the assumption of the Depth of Discharge (DoD) BESS of 0.1. When the DoD of BESS is less than 0.1, the shunt inverter will act as a load to the grid. Therefore, the shunt inverter absorbs the power from the grid and delivers it to the series inverter and the series inverter compensates for the load voltage as shown in Figure 18c. The control of the shunt inverter will be transferred to the load mode. The power absorbed from the shunt inverter will be delivered to the DC capacitor to keep the voltage constant. In Figure 18b, it is shown that during t > 2 s the undervoltage is worse due to the additional load from the shunt inverter. The loads will increase because of the shunt inverter as an additional load and due to that the undervoltage will increase too. That is why the series inverter injects more voltage to compensate for the grid voltage after t > 2 s.

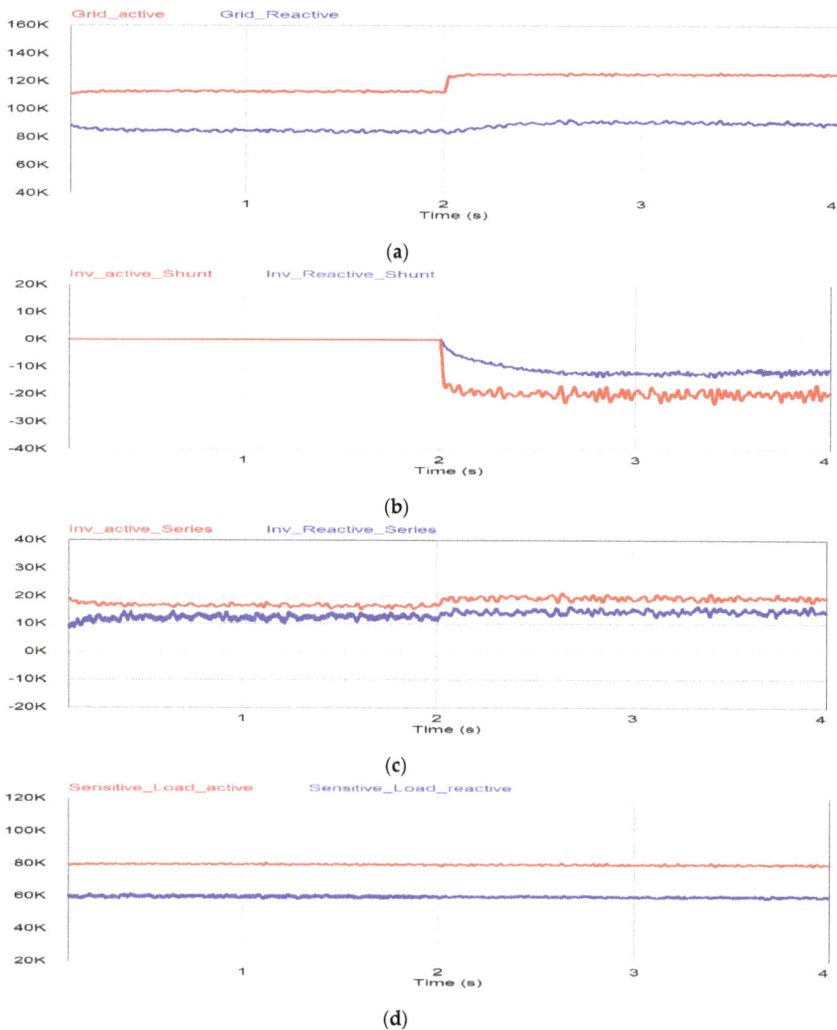

Figure 18. *Cont.*

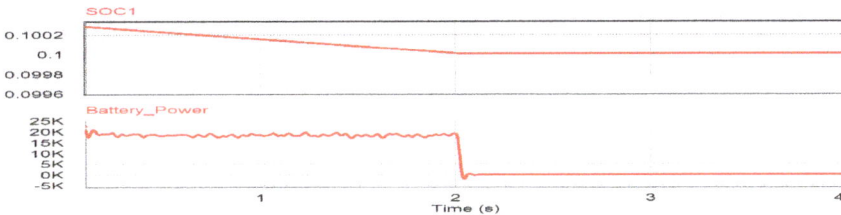

(e)

Figure 18. Simulation results of voltage compensation from grid power mode: (**a**) Grid power; (**b**) shunt inverter power; (**c**) series inverter power; (**d**) active and reactive power for sensitive load; (**e**) state of charge of BESS and BESS power.

6. Conclusions

This paper proposes multi-functional PV inverters based on the UPQC configuration in which the back-to-back inverters are connected to the grid by series and shunt. The line voltage is compensated for by the series inverter, while the shunt inverter delivers the PV generating power to the grid. From the point of view of voltage compensation, it has been compared to the conventional DVR, which can be divided by the external voltage compensator for the distributed generating system. The UPQC configuration has been used in place of the conventional grid-connected inverter and the external DVR for the compensation of voltage while delivering the PV generating power. Finally, the microgrid concept has been applied to keep the different power quality levels depending on the importance of the loads.

Depending on the grid voltage and the condition of weather and battery SOC, it can operate in one of five different modes: normal mode, voltage compensation mode, stand-alone mode, voltage compensation mode from BESS and voltage compensation mode from grid. The operation performance at each mode is analyzed by using the PSiM simulation with the voltage waveforms and the active and reactive power flows. From the simulation results, the proposed idea is well verified so that it can be applied for the power quality improvement of a weak grid power system such as in isolated areas or on islands connected to the mainland by long submarine cables.

Author Contributions: Conceptualization and Methodology, D.P.S. and J.C.; Software, Validation, and Writing-Original Draft Preparation, D.P.S.; Writing-Review, Editing and Supervision, J.C.

Acknowledgments: This work was supported by the Korea Institute of Energy Technology Evaluation and Planning (KETEP) and the Ministry of Trade, Industry & Energy (MOTIE) of the Republic of Korea (No. 20171210200840).

Conflicts of Interest: The authors declare no conflict of interest.

References

1. United Nations. *Kyoto Protocol to the United Nations Framework Convention on Climate Change*; United Nations: New York, NY, USA, 1998.
2. IEA International Energy Agency. Snapshot of Global Photovoltaic Markets: Report IEA PVPS T1-33. 2018. Available online: http://www.iea-pvps.org/fileadmin/dam/public/report/statistics/IEA-PVPS_-_A_Snapshot_of_Global_PV_-_1992-2017.pdf (accessed on 8 July 2018).
3. *Guide for Planning DC Links Terminating at AC Systems Locations Having Low Short-Circuit Capacities, Part I: AC/DC Interaction Phenomena*; IEEE Std. 1204-1997; IEEE: Piscataway, NJ, USA, 1997.
4. Keyhani, A. *Design of Smart Power Grid Renewable Energy Systems*, 2nd ed.; Wiley-IEEE Press: Piscataway, NJ, USA, 2015.
5. *IEEE Recommended Practice for Monitoring Electric Power Quality*; IEEE Std. 1159-2009; IEEE: Piscataway, NJ, USA, 1999.

6. Nagpal, M.; Martinich, T.G.; Bimbhra, A.; Sydor, D. Damaging open-phase overvoltage disturbance on a shunt-compensated 500-kV line initiated by unintended trip. *IEEE Trans. Power Deliv.* **2015**, *30*, 412–419. [CrossRef]

7. Chavan, G.; Acharya, S.; Bhattacharya, S.; Das, D.; Imam, H. Application of static synchronous series compensators in mitigating Ferranti effect. In Proceedings of the 2016 IEEE Power and Energy Society General Meeting (PESGM), Boston, MA, USA, 1–5 July 2016.

8. Natesan, C.; Ajithan, K.; Palani, P.; Kandhasamy, P. Survey on microgrid: Power quality improvement techniques. *Renew. Energy* **2014**, 342019. [CrossRef]

9. AppalaNaidu, T. The Role of Dynamic Voltage Restorer (DVR) in improving power quality. In Proceedings of the 2016 International Conference on Advances in Electrical, Electronics, Information, Communication and Bio-Informatics (AEEICB), Chennai, India, 27–28 February 2016.

10. Fernandes, D.; Costa, F.; Martins, J.; Lock, A.; Silva, E.; Vitorino, M. Sensitive load voltage compensation Performed by a Suitable Control Method. *IEEE Trans. Ind. Appl.* **2017**, *53*, 4877–4885. [CrossRef]

11. Sin, C.; Choi, J. Compensation of current harmonics caused by local nonlinear load for grid-connected converter. In Proceedings of the IFEEC 2017-ECCE Asia, Kaohshiung, Taiwan, 3–7 June 2017.

12. Jayaprakash, P.; Singh, B.; Kothari, P.; Chandra, A.; Al-Haddad, K. Control of reduced-rating dynamic voltage restorer with a battery energy storage system. *IEEE Trans. Ind. Appl.* **2014**, *50*, 1295–1303. [CrossRef]

13. Gee, A.; Robinson, F.; Yuan, W. A Superconducting magnetic energy storage-emulator/battery supported dynamic voltage restorer. *IEEE Trans. Energy Convers.* **2017**, *32*, 55–64. [CrossRef]

14. Choi, S.; Li, J.; Vilathgamuwa, D. A generalized voltage compensation strategy for mitigating the impacts of voltage sags/swells. *IEEE Trans. Power Deliv.* **2005**, *20*, 2289–2297. [CrossRef]

15. Khadkikar, V. Enhancing electric power quality using UPQC: A comprehensive overview. *IEEE Trans. Power Electron.* **2012**, *27*, 2284–2297. [CrossRef]

16. Graovac, D.; Katic, V.; Rufer, A. Power quality compensation using universal power quality conditioning system. *IEEE Power Eng. Rev.* **2000**, *20*, 58–60.

17. Somayajula, D.; Crow, M. An integrated dynamic voltage restorer-ultracapacitor design for improving power quality of the distribution grid. *IEEE Trans. Sustain. Energy* **2015**, *6*, 616–624. [CrossRef]

18. Karimian, M.; Jalilian, A. Proportional-repetitive control of a dynamic voltage restorer (DVR) for power quality improvement. In Proceedings of the 17th Conference Electrical Power Distribution, Tehran, Iran, 1–6 May 2012.

19. Fernandes, D.; Costa, F.; Martins, J.; Lock, A.; Silva, E.; Vitorino, M. Sensitive load voltage compensation with a suitable control method. In Proceedings of the IEEE Energy Conversion Congress Exposition, Montreal, QC, Canada, 20–24 September 2015.

20. Balaguer, I.; Lei, Q.; Yang, S.; Supatti, U.; Peng, F. Control for grid-connected and intentional islanding operations of distributed power generation. *IEEE Trans. Ind. Electron.* **2011**, *58*, 147–157. [CrossRef]

21. Ramadesigan, V.; Northrop, P.; De, S.; Santhanagopalan, S.; Braarz, R.; Subramanian, V. Modeling and simulation of lithium-ion batteries from a system engineering perspective. *J. Electrochem. Soc.* **2012**, *159*, R31–R45. [CrossRef]

22. Domenico, D.; Creef, Y.; Prada, E.; Duchene, P.; Bernard, J.; Sauvan-Moynot, V. A review of approaches for the design of Li-ion BMS estimation functions. *J. Oil Gas Sci. Technol.* **2013**, *68*, 127–135. [CrossRef]

23. Grainger, J.; William, D.S. *Power System Analysis*; McGraw-Hill Inc.: New York, NY, USA, 1994.

24. Song-Manguelle, J.; Harfman, T.M.; Chi, S.; Gunturi, S.K.; Datta, R. Power transfer capability of HVAC cable for subsea transmission and distribution system. *IEEE Trans. Ind. Appl.* **2014**, *50*, 2382–2391. [CrossRef]

25. Jothibasu, S.; Mishra, M.A. Control scheme for storageless DVR based on characterization of voltage sags. *IEEE Trans. Power Deliv.* **2014**, *29*, 2261–2269. [CrossRef]

26. Nielsen, J.G.; Blaabjerg, F. A detailed comparison of system topologies for dynamic voltage restorers. *IEEE Trans. Ind. Appl.* **2015**, *41*, 1272–1280. [CrossRef]

27. Woodley, N.H.; Morgan, L.; Sundaram, A. Experience with an inverter-based dynamic voltage restorer. *IEEE Trans. Power Deliv.* **1999**, *14*, 1181–1186. [CrossRef]

28. Sanchez, P.; Acha, E.; Calderon, J.; Feliu, V.; Cerrada, A. A versatile control scheme for a dynamic voltage restorer for power quality improvement. *IEEE Trans. Power Deliv.* **2009**, *24*, 277–284. [CrossRef]

29. Lam, C.S.; Wong, M.C.; Han, Y.D. Voltage swell and overvoltage compensation with unidirectional power flow controlled dynamic voltage restorer. *IEEE Trans. Power Deliv.* **2008**, *23*, 2513–2521.

30. Sadigh, A.; Smedley, K. Review of voltage compensation methods in dynamic voltage restorer DVR. In Proceedings of the IEEE Power and Energy Society General Meeting, San Diego, CA, USA, 22–26 July 2012.

31. Rauf, A.M.; Khadkikar, V. An enhanced voltage sag compensation scheme for dynamic voltage restorer. *IEEE Trans. Ind. Electron.* **2015**, *62*, 2683–2692. [CrossRef]

![energies logo] *energies*

MDPI

Article

Real-Time Implementation of Robust Control Strategies Based on Sliding Mode Control for Standalone Microgrids Supplying Non-Linear Loads

Seghir Benhalima, Rezkallah Miloud * and Ambrish Chandra

École de Technologie Supérieure, University of Quebec, Montreal, QC H3C 1K3, Canada;
seghir.benhalima.1@ens.etsmtl.ca (S.B.); ambrish.chandra@etsmtl.ca (A.C.)
* Correspondence: miloud.rezkallah.1@ens.etsmtl.ca; Tel.: +1-514-754-7315

Received: 31 July 2018; Accepted: 24 September 2018; Published: 28 September 2018

Abstract: In this paper enhanced control strategies for standalone microgrids based on solar photovoltaic systems (SVPAs) and diesel engine driven fixed speed synchronous generators, are presented. Single-phase d-q theory-based sliding mode controller for voltage source converter voltage source converter (VSC) is employed to mitigate harmonics, balance diesel generator (DG) current, and to inject the generated power by SVPA into local grid. To achieve fast dynamic response with zero steady-state error during transition, sliding mode controller for inner control loop is employed. To achieve maximum power point tracking (MPPT) from SVPA without using any MPPT method, a DC-DC buck boost converter supported by battery storage system is controlled using a new control strategy based on sliding mode control with boundary layer. In addition, modeling and detailed stability analysis are performed. The performance of the developed control strategies, are validate by simulation using MATLAB/Simulink and in real-time using hardware prototype.

Keywords: standalone microgrid; sliding mode control; solar photovoltaic system; diesel generator; power quality improvement; stability analysis

---

1. Introduction

Most of remote areas in the world are not connected to the main grid. Generally, conventional diesel generators (DGs) are employed as the energy source to provide electricity to the connected loads in these areas [1–3]. Considering the transport and fuel costs, electricity in these isolated areas is costly. In addition, the performance of a DG is reduced when it is operated at light load, as well as, when connected to balanced and unbalanced nonlinear loads [4]. Many solutions are proposed in the literature to solve these issues and to reduce CO_2 emissions [5–8]. In [9], a hybrid wind-diesel system-based solution is proposed, and in [10,11] solar-wind-diesel hybrid systems are suggested. These solutions are effective from the point of view of fuel consumption but hard to implement in practice. Recently, in [12], many technical solutions with reduced number of power converters, are suggested to reduce the complexity of hybrid standalone system and achieve high performance from the available energy sources (ESs) in isolated localities.

Solar is the most abundant renewable energy source (RES) in the world. This RES is stochastic, not available at night, and it cannot dispatch power directly to the load. To ensure uninterruptible power supply to connected isolated loads, DG and storage elements, such as batteries, are required as reliable ESs. In a standalone system configuration proposed in [13], a solar photovoltaic system (SPVA) is connected to the DC bus through a DC-DC boost converter and a battery is connected directly to the DC bus without using any power converter. However, in [14], SPVA and batteries are connected to the DC bus through power converters. This configuration is costly and hard to implement in a real situation. In addition, operating many power converters, which are connected to the same

DC bus using different switching frequency is challenging and can damage the batteries, and affect the system stability. In [13,14], the DC-DC boost converter and the DC-DC buck boost converter, are controlled using simple control algorithms based on a linear proportional integral (PI) controller, which is not robust and cannot perform well under large variations of the system parameters or loading conditions. To overcome these problems and achieve high performance without saturation issues, sliding mode control is proposed in [15]. The obtained simulation results showed satisfactory performance. Unfortunately, in [15] the chattering issue is not considered. To overcome this issue, terminal sliding mode control (SMC), and high-order SMC, are proposed in [16,17]. The obtained experimentally results show satisfactory performance without any chattering issues. Compared to the classical SMC proposed in [15], terminal and high-order SMCs are robust but complex and require more compilation time, which is hard to implement using simple microprocessors.

Generally, in the conventional grid, power quality issues at the point of common coupling (PCC), such as voltage sag and swell, harmonics, etc. can be solved using filters [18]. This solution is also adopted in standalone microgrid to improve the power quality at the point of common coupling (PCC).

In standalone microgrids based on SPVA and DG, the interfacing power electronic converter is mostly controlled to achieve various tasks, such as regulation of the voltage and frequency, as well as improve the power quality at PCC [19,20]. Regarding the harmonics mitigation and to balance the source current at PCC, many control strategies are proposed in the literature such as the synchronous reference frame (SRF) control strategy applied in [21], or the instantaneous p-q theory developed first by Akagi and applied to improve the power quality in standalone microgrid in [18]. In [18], direct and indirect controls are suggested. In the control strategies proposed in [18,21], the authors have used a conventional proportional integral (PI) controller to track the errors in outer and inner loops. This controller is simple but requires tuning to achieve optimal regulation, which is hard to achieve, especially if the proposed control strategy requires more than one PI controller. Furthermore, PI controllers can saturate during transition and when input signals are polluted. To solve this issue, a proportional integral based artificial neural network fuzzy interference system (PI-ANFIS) controller is proposed in [22], and in [23] an adaptive controller is suggested. These controllers are robust compared to simple PI controllers but require more computation time, which is hard to implement in real time using simple microprocessor chips. To overcome the issue of the linear PI controller, a fuzzy sliding mode controller is proposed in [24], in [25] a neural global SMC using a fuzzy approximator is employed, and in [26], an adaptive fractional fuzzy SMC, is suggested. The solutions proposed in [24–26], are effective but complex to implement, especially if the inner and outer loop control requires more than one controller. Sliding mode (SM)-based control is proposed in [27] for the inner loop control in order instead of the conventional PI controller to ensure a fast-time and optimal tracking control. The obtained experimental and simulation results show satisfactory performance. Unfortunately, the proposed concept is dedicated to controlling an interfacing voltage source converter, which is connected to the grid, where the frequency and voltage regulation are not achieved by the control system, and synchronization between the interfacing converter and the grid is not a big challenge. In addition, DC link voltage regulation is ensured by a PI controller, where saturation issues are not considered.

To solve these issues, which are related to the cost of energy in isolated areas, power quality and stable operation of standalone microgrids based on SPVA and DG, the following solutions are proposed:

(1) Standalone microgrid -microgrid configurations with less power converters where SPVA and battery energy storage system are controlled using only one DC-DC buck-boost converter to connect and control the lead acid battery pack.

(2) Using only one PI controller with anti-windup for the outer control loop for DC-link voltage regulation.

(3) Replacing the PI controller by a SMC with boundary layer in the inner control loop for single-phase d-q control strategy, which is proposed for power quality improvement.

(4) Developing a new control strategy based on sliding mode control for DC-DC buck boost converter where the system parameters are taken on consideration in equivalent control.

(5) Achieving maximum power point tracking (MPPT) from SPVA by controlling only the DC link voltage.

(6) Employing the boundary layer in SMC to reduce the chattering phenomena.

2. System Description and Operation Mode

Figure 1 shows a standalone microgrid-based configuration for remote areas. It consists of two energy sources: (1) a fixed speed diesel engine driven synchronous generator, and (2) a SPVA. The DG is connected directly to the PCC and the SPVA is connected directly to the common DC bus and to the PCC through a three-phase voltage source converter (VSC). A lead acid battery pack is connected to the common DC bus through a controller DC-DC buck boost converter. For the proposed configuration, the AC voltage and system frequency are regulated by the governor and automatic voltage regulator (AVR) of the DG. For power quality improvement at the PCC and injection of the generated power from SPVA, the VSC is controlled using a single-phase d-q theory based-sliding mode control strategy. Regarding the power balance in the system and for achieving MPPT from SPVA, an enhanced control strategy based on sliding mode control is used to control the DC-DC buck boost converter. To reduce the ripple voltage due to high switching frequency of VSC, a resistance-inductor-capacitor (RLC) passive filter is employed.

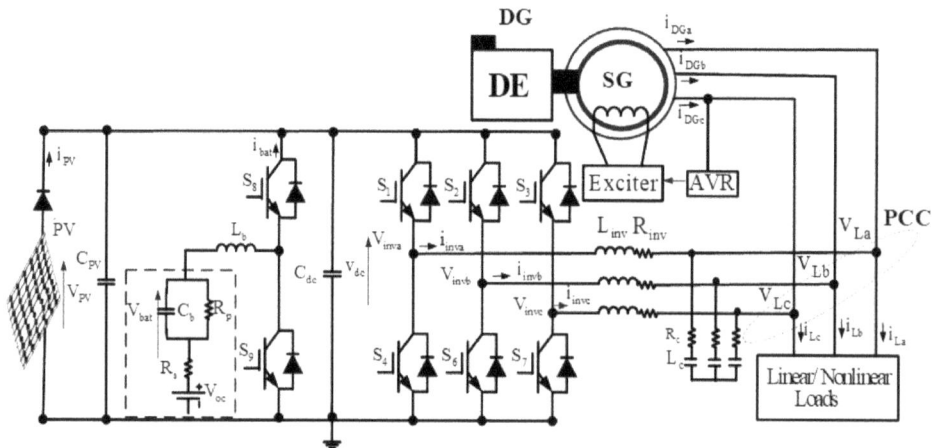

Figure 1. The standalone microgrid configuration under study.

3. Control System

In this section, the developed control strategies for the DC-DC buck boost converter and for three-phase voltage source converter (VSC), are detailed.

3.1. Control of the DC-DC Buck-Boost Converter

As shown in Figure 1 the lead acid battery is connected to the common DC-link through a controlled DC-DC buck-boost converter, which consists of two continuous systems. To balance the power in the system and achieve MPPT from SPVA, sliding mode control (SMC) with boundary layer is used.

3.1.1. Modeling of the DC-DC Buck-Boost Converter

The DC-DC buck-boost converter shown in Figure 1 consists of battery voltage (V_{bat}) controlled switches (S_8 and S_9), inductor (L_b), output voltage (V_{dc}) and a capacitor (C_{dc}). Depending on the state of the switches, the state-space equations of the DC-DC buck boost converter when the switch is ON are written as:

$$\begin{cases} \frac{di_{bat}}{dt} = \frac{V_{bat}}{L_b} \\ \frac{dV_{dc}}{dt} = -\frac{V_{dc}}{RC_{dc}} \end{cases} \tag{1}$$

where R is load resistance, and when the switch is OFF the state-space equations are expressed as:

$$\begin{cases} \frac{di_{bat}}{dt} = \frac{V_{dc}}{L_b} \\ \frac{dV_{dc}}{dt} = -\frac{i_{bat}}{C_{dc}} - \frac{V_{dc}}{RC_{dc}} \end{cases} \tag{2}$$

Replacing, x_1 by i_{bat}, and x_2 by V_{dc}, then; $\dot{x}_1 = \frac{di_{bat}}{dt}$, $\dot{x}_2 = \frac{dV_{dc}}{dt}$. The state-space equations, which are expressed in (1) and (2) are written as:

$$\begin{cases} \dot{x}_1 = A_1 x + B_1 u \\ V_{dc} = C_1 x \end{cases} \tag{3}$$

where $A_1 = \begin{bmatrix} 0 & 0 \\ 0 & -\frac{1}{RC_{dc}} \end{bmatrix}$, $B_1 = \begin{bmatrix} \frac{1}{L_b} \\ 0 \end{bmatrix}$, $C_1 = \begin{bmatrix} 0 & 1 \end{bmatrix}$ and $u = v_{bat}$

$$\begin{cases} \dot{x}_2 = A_2 x + B_2 u \\ V_{dc} = C_2 x \end{cases} \tag{4}$$

where $A_2 = \begin{bmatrix} 0 & \frac{1}{L_b} \\ -\frac{1}{C_{dc}} & -\frac{1}{RC_{dc}} \end{bmatrix}$, $B_1 = \begin{bmatrix} 0 \\ 0 \end{bmatrix}$, $C_1 = \begin{bmatrix} 0 & 1 \end{bmatrix}$ and $u = v_{bat}$

The average state-space equation of the DC-DC buck-boost converter is obtained using (3) and (4) as,

$$\begin{cases} \dot{x}(t) = Ax(t) + Bu(t) \\ V_{dc} = Cx(t) \end{cases} \tag{5}$$

where $A = \begin{bmatrix} 0 & \frac{1-d}{L_b} \\ -\frac{1-d}{C_{dc}} & -\frac{1}{RC_{dc}} \end{bmatrix}$, $B = \begin{bmatrix} \frac{d}{L_b} \\ 0 \end{bmatrix} v_{bat}$, and $C = \begin{bmatrix} 0 & 1 \end{bmatrix}$.

3.1.2. Sliding Mode Control for the DC-DC Buck-Boost Converter

Figure 2 shows the block diagram of the developed control strategy for the DC-DC buck boost converter. Based on (6), the model of the DC-DC buck-boost converter is nonlinear. However, for achieving high dynamic performance during transition with optimal accuracy regulation and robust trajectory tracking in presence of uncertain parameters, the SMC approach is employed. The control design of the general control (d) for the DC-DC buck boost converter is obtained using the following steps:

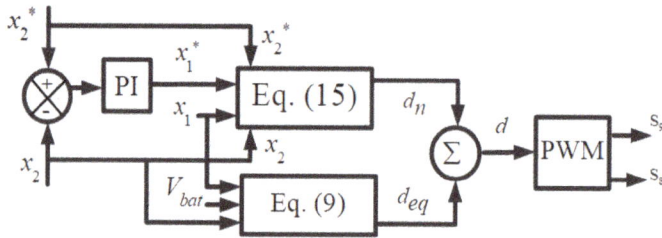

Figure 2. Control of the DC-DC buck boost converter.

3.1.3. Selection of the Sliding Surface

The sliding surface (σ) as expressed in (6) is selected to ensure reaching the surface with desiring dynamics of the corresponding sliding motion [28–30]:

$$\sigma = \beta_1(x_1 - x_1^*) + \beta_2(x_2 - x_2^*) \tag{6}$$

where β_1 and β_2 represent the sliding gains and they should be positive.

The x_1^* and x_2^* denote the battery current and DC-link voltage references, respectively. The value of the DC-link voltage reference is equal to output PV voltage (V_{PV}), which is selected equal to 350 V. As presented in Figure 3, V_{PV} corresponds to the maximum extracted power from the SPVA for different solar irradiations. Therefore, by controlling the DC-link voltage, one can easily extract the maximum of power without using any MPPT method.

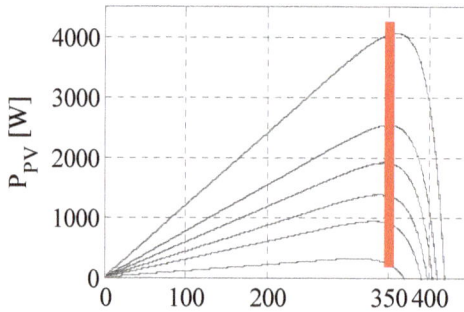

Figure 3. $P_{PV} = f(V_{PV})$ with fixed temperature and solar irradiation change.

3.1.4. Equivalent Control

The structure of the desired control (d) is obtained as:

$$d = d_{eq} + d_n \tag{7}$$

where d_n represents the nonlinear control and d_{eq} is the equivalent control, which is obtained using (5), (6) and by setting the derivative of (6) to zero as:

$$\dot{\sigma} = 0 \tag{8}$$

By replacing $\dot{x}_1$ and $\dot{x}_2$ by their equality, one obtains the following expression as:

$$d_{eq} = \frac{x_2(\beta_1 RC_{dc} - \beta_2 L_b) - \beta_2 RL_b x_1}{\beta_2 RL_b x_1 + \beta_1 RC_{dc}(x_2 - V_{bat})} \tag{9}$$

3.1.5. Stability Analysis

The objective of the approach using SMC is to ensure the convergence of the operating points to define the sliding boundary. Therefore, for assuring the stability of the control, Lyapunov stability function. According to Lyapunov theorem, a non-linear time variant system is globally and uniformly stable if satisfies the following conditions as [31,32]:

$$
\begin{cases}
V(0) = 0 \\
V(X) > 0 \\
V(X) < 0 \\
\text{Then } X = 0 \text{ is asymptotically stable}
\end{cases}
\tag{10}
$$

For the DC-DC buck-boost converter Lyapunov function is defined as [33,34] as:

$$
V = \frac{1}{2}\sigma^2
\tag{11}
$$

The system is considered asymptotically stable as detailed in (10), the derivative of (11) should be negative:

$$
\dot{V} = \sigma\dot{\sigma} < 0
\tag{12}
$$

Replacing (6) and (8) in (12), one gets:

$$
\dot{V} = \sigma\dot{\sigma} = (\beta_1(x_1 - x_1^*) + \beta_2(x_2 - x_2^*))(\beta_1\dot{x}_1 + \beta_2\dot{x}_2) < 0
\tag{13}
$$

where the x_1^* and x_2^* denote the references of the battery current, which represents the output of the PI controller with anti-windup DC link voltage regulator, and DC-link voltage, which is equal to the 350 V, respectively

Substituting terms $\dot{x}_1$, $\dot{x}_2$, x_1, x_2, and d by their equivalences in (12) gives the following expression:

$$
\dot{V} = \sigma\dot{\sigma} = -
\begin{bmatrix}
\overbrace{\left(\frac{V_{dc}^2\beta_2^2}{RC_{dc}}\right) + \left(\frac{i_{bat}^2 V_{bat}\beta_2\beta_1}{V_{dc}}\right) + \left(\frac{V_{bat}^2\beta_2\beta_1}{L_b}\right) + \left(\frac{V_{bat}\beta_2\beta_1}{L_b}x_2^*\right)}^{Term1} \\
+ \left(\frac{V_{dc}i_{bat}\beta_2\beta_1}{RC_{dc}}\right) + \left(V_{dc}i_{bat}\beta_2^2 x_1^*\right)
\end{bmatrix}
$$
$$
+
\begin{bmatrix}
\overbrace{\left(\frac{V_{dc}^2\beta_1}{L_b}\right) + \left(V_{bat}\beta_1^2 i_{bat}\right) + \left(\frac{V_{dc}\beta_2}{RC_{dc}}\right) + \left(\frac{V_{dc}\beta_2\beta_1}{RC_{dc}}x_1^*\right) + \left(\frac{V_{bat}V_{dc}\beta_2\beta_1}{L_b}\right)}^{Term2} \\
+ \left(\frac{V_{bat}\beta_1^2 i_{bat}}{L_b}\right) + \left(\frac{V_{bat}\beta_1^2 i_{bat}}{C_{dc}V_{dc}}x_2^*\right) + \left(\frac{i_{bat}V_{bat}\beta_2\beta_1}{C_{dc}V_{dc}}x_1^*\right)
\end{bmatrix}
< 0
\tag{14}
$$

Replacing the parameters L_b, C_{dc}, V_{bat}, V_{dc}, as well as the optimal gains of the SMC β_1 and β_2 by their value given in Table 1, the first term with negative sign in (14) is larger than the second term. This leads that condition given in (13) is satisfied, which confirms that the system is asymptotically stable.

Table 1. System parameters.

Parameters	Value	Parameters	Value
v_{DG}	208 V	k_1, k_{i1},	0.1, 25
f_r	60 Hz	F_{sw} VSC	10 kHz
V_{dc}^*	350 V	C_{dc}	1000 μF
C_{PV}	100 μF	L_{inv}	5 mH
L_b	1.5 mH	R_{inv}	0.01 Ω
F_{sw} DC-DC Converter	10 kHz	R_c	5 Ω
β_1, β_2 and β_3	0.001, 0.8 and 5	C_c	10 μF

To ensure the robustness, the second of the desired control d_n, which represents the discontinuous control is written as [35–37]:

$$d_n = \beta_3 sat(\sigma, \varnothing) \tag{15}$$

where β_3 is positive control gain.

To reduce the chattering of the control signal, saturation function with boundary conditions is defined as:

$$sat(\sigma, \varnothing) \begin{cases} 1 & \sigma > \varnothing \\ \frac{\sigma}{\varnothing} & |\sigma| \leq \varnothing \\ -1 & \sigma < -\varnothing \end{cases} \tag{16}$$

where $\varnothing$ denotes the sliding layer, which is selected between 0.5 and -0.5 [11].

The optimal gains (β_1, β_2 and β_3) of the SMC, are selected equal to $\beta_1 = 0.001$, $\beta_2 = 0.8$, and $\beta_3 = 5$, respectively. One observes in Figure 4c, that based on the optimal values of β_1, β_2 and β_3 the signal reaches the origin rapidly.

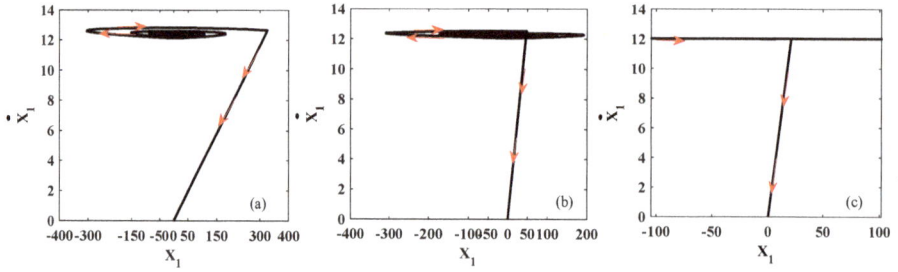

Figure 4. Diagram of $\dot{x}_1 = f(x_1)$ with: (a) $\beta_1 = 0.001$, $\beta_2 = 0.8$, and $\beta_3 = 0.05$, (b) $\beta_1 = 0.01$, $\beta_2 = 0.8$, and $\beta_3 = 0.09$, and (c) $\beta_1 = 0.001$, $\beta_2 = 0.8$, and $\beta_3 = 5$.

3.2. Control of Three-Phase Voltage Source Converter

In Figure 5 the single-phase d-q theory-based sliding mode controller for VSC to improve the power quality and inject the generated power from the solar photovoltaic system into PCC, is presented. Regarding the DC link voltage regulation, is ensured by controlling the DC-DC buck boost converter. Inverter voltage (v_{invabc}) and current (i_{invabc}), load current (i_{Labc}) and voltage(v_{Labc}), are used to determine the inverter voltage references (v^*_{invabc}). The d-q rotating frame and (α-β) stationary frame are employed to transform all time-varying signals to DC quantities with time-invariant.

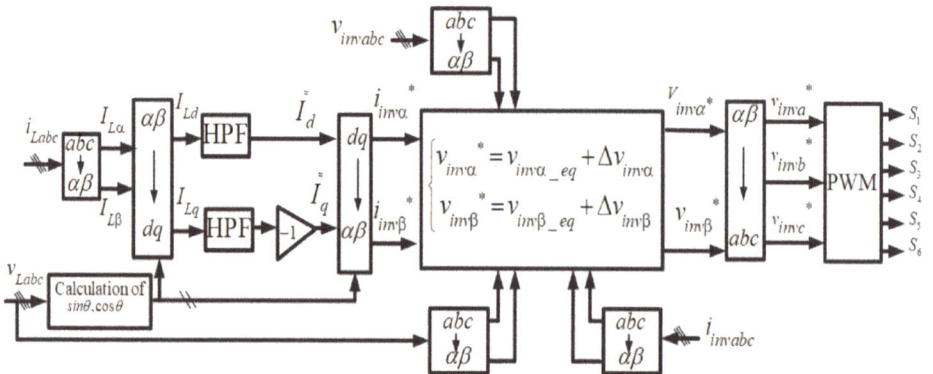

Figure 5. Single-phased d-q theory-based on SMC.

The rotating angle of d-q frame is calculated using the in-phase and quadrature united vectors as in [18]. The amplitude of the PCC voltage (V_p) is calculated as:

$$V_p = \sqrt{\frac{2}{3}\left((v_{La}^2 + v_{Lb}^2 + v_{Lc}^2)\right)} \tag{17}$$

The in-phase unit templates of the PCC voltage, are calaculted as follows:

$$u_{pa} = \frac{v_{La}}{V_p}, u_{pb} = \frac{v_{Lb}}{V_p}, u_{pc} = \frac{v_{Lc}}{V_p} \tag{18}$$

and the quadrature unit templates of the PCC voltages are expreesed as:

$$u_{qa} = -\frac{u_{pa}}{\sqrt{3}} + \frac{u_{pc}}{\sqrt{3}}, u_{qb} = \sqrt{3}\frac{u_{pa}}{2} + \frac{\left(u_{pb} - u_{pc}\right)}{2\sqrt{3}}, u_{qc} = -\sqrt{3}\frac{u_{pa}}{2} + \frac{\left(u_{pb} - u_{pc}\right)}{2\sqrt{3}} \tag{19}$$

The $cos\theta$ and $sin\theta$, are defined as:

$$\begin{cases} sin\theta = u_{pa} \\ cos\theta = u_{qa} \end{cases} \tag{20}$$

The obtained dq-axis load current consist of the fundamentals ($\overline{I_d}$, $\overline{I_q}$) and harmonics components ($\tilde{I_q}$, $\tilde{I_d}$):

$$\begin{cases} I_{Ld} = \overline{I_d} + \tilde{I_d} \\ I_{Lq} = \overline{I_q} + \tilde{I_q} \end{cases} \tag{21}$$

Regarding the harmonics ($\tilde{I_q}$, $\tilde{I_d}$), which are extracted using high pass filter (HPF) represent the output inverter currents, they are transformed into the (α-β) stationary frame., and are used after to determine the reference inverter voltages.

3.2.1. Sliding Mode Current Controller

A sliding mode current controller (SMCC) is proposed for the inner current control loop to achieve high performance during transition with a fast-dynamic response. To design the controller, the following steps are used.

3.2.2. Selecting the Switching Surface

Switching surface for SMCC $\sigma_1 = \begin{bmatrix} \sigma_\alpha & \sigma_\beta \end{bmatrix}^T$ is selected to ensure fast dynamic response during transition with zero steady-state [38]:

$$\begin{cases} \sigma_\alpha = k_1 e_\alpha(t) + k_{i1} \int\limits_0^t e_\alpha(i_{inv\alpha})dt \\ \sigma_\beta = k_1 e_\beta(t) + k_{i1} \int\limits_0^t e_\beta(i_{inv\beta})dt \end{cases} \tag{22}$$

where k_1, and k_{i1} are positive gains. The e_α, e_β represent the obtained errors between measured and reference output VSC currents, which are defined as:

$$\begin{cases} e_\alpha = \left(i_{inv\alpha}^* - i_{inv\alpha}\right) \\ e_\beta = \left(i_{inv\beta}^* - i_{inv\beta}\right) \end{cases} \tag{23}$$

Applying the Kirchhoff current and voltage law at the output of the VSC, one obtains the following expression in stationary (α-β) frame:

$$\begin{cases} \dot{i}_{inv\alpha} = \frac{1}{L_{inv}}(v_{inv\alpha} - v_{L\alpha}) - \frac{R_{inv}}{L_{inv}}i_{inv\alpha} \\ \dot{i}_{inv\beta} = \frac{1}{L_{inv}}(v_{inv\beta} - v_{L\beta}) - \frac{R_{inv}}{L_{inv}}i_{inv\beta} \end{cases} \tag{24}$$

where L_{inv} and R_{inv} denote the inductor and resistor of the output filter of VSC.

Using (24), (22) and (23) and its derivative, one obtains the following expression:

$$\begin{cases} \dot{\sigma}_{\alpha} = k_1\left(\dot{i}^*_{inv\alpha} - \left(\frac{1}{L_{inv}}(v_{inv\alpha} - v_{L\alpha}) - \frac{R_{inv}}{L_{inv}}i_{inv\alpha}\right)\right) + k_{i1}e_{\alpha} \\ \dot{\sigma}_{\beta} = k_1\left(\dot{i}^*_{inv\beta} - \left(\frac{1}{L_{inv}}(v_{inv\beta} - v_{L\beta}) - \frac{R_{inv}}{L_{inv}}i_{inv\beta}\right)\right) + k_{i1}e_{\beta} \end{cases} \tag{25}$$

The structure of the desired control ($v^*_{inv\alpha}$, and $v^*_{inv\beta}$), are expressed as:

$$\begin{cases} v^*_{inv\alpha} = v_{inv\alpha\_eq} + \Delta v_{inv\alpha} \\ v^*_{inv\beta} = v_{inv\beta\_eq} + \Delta v_{inv\beta} \end{cases} \tag{26}$$

where $v_{inv\alpha\_eq}$, $v_{inv\beta\_eq}$, $\Delta v_{inv\alpha}$, and $\Delta v_{inv\beta}$ represent the equivalent controls and nonlinear control in the stationary (α-β) frame.

Regarding, the equivalent controls are obtained by set the sliding switching $\dot{\sigma}_1 = \begin{bmatrix} \dot{\sigma}_{\alpha} & \dot{\sigma}_{\beta} \end{bmatrix}^T$ equal to zero:

$$\begin{cases} \dot{\sigma}_{\alpha} = k_1\left(\dot{i}^*_{inv\alpha} - \left(\frac{1}{L_{inv}}(v_{inv\alpha\_eq} - v_{L\alpha}) - \frac{R_{inv}}{L_{inv}}i_{inv\alpha}\right)\right) + k_{i1}e_{\alpha} = 0 \\ \dot{\sigma}_{\beta} = k_1\left(\dot{i}^*_{inv\beta} - \left(\frac{1}{L_{inv}}(v_{inv\beta\_eq} - v_{L\beta}) - \frac{R_{inv}}{L_{inv}}i_{inv\beta}\right)\right) + k_{i1}e_{\beta} = 0 \end{cases} \tag{27}$$

From (25), one obtains the equivalent controls stationary (α-β) frame:

$$\begin{cases} v_{inv\alpha\_eq} = \frac{k_{i1}}{k_1}L_{inv}e_{\alpha} + v_{L\alpha} + R_{inv}i_{inv\alpha} + L_{inv}\dot{i}^*_{inv\alpha} \\ v_{inv\beta\_eq} = \frac{k_{i1}}{k_1}L_{inv}e_{\beta} + v_{L\alpha} + R_{inv}i_{inv\beta} + L_{inv}\dot{i}^*_{inv\beta} \end{cases} \tag{28}$$

The nonlinear control is selected equal to:

$$\begin{cases} \Delta v_{inv\alpha} = K_2 Sat(\sigma_{\alpha}) \\ \Delta v_{inv\beta} = K_2 Sat(\sigma_{\beta}) \end{cases} \tag{29}$$

where K_2 represents the controller gain and is positive.

3.2.3. Stability Analysis

The system is considered globally stable only if the derivative of (10) is negative. Using (26) and (22), one gets the following expression:

$$\begin{cases} \sigma_{\alpha}\dot{\sigma}_{\alpha} = \left(k_1\left(\dot{i}^*_{inv\alpha} - \left(\frac{1}{L_{inv}}(v_{inv\alpha} - v_{L\alpha}) - \frac{R_{inv}}{L_{inv}}i_{inv\alpha}\right)\right) + k_{i1}e_{\alpha}\right)\left(k_1e_{\alpha}(t) + k_{i1}\int_0^t e_{\alpha}(i_{inv\alpha})dt\right) < 0 \\ \sigma_{\beta}\dot{\sigma}_{\beta} = \left(k_1\left(\dot{i}^*_{inv\beta} - \left(\frac{1}{L_{inv}}(v_{inv\beta} - v_{L\beta}) - \frac{R_{inv}}{L_{inv}}i_{inv\beta}\right)\right) + k_{i1}e_{\beta}\right)\left(k_1e_{\beta}(t) + k_{i1}\int_0^t e_{\beta}(i_{inv\beta})dt\right) < 0 \end{cases} \tag{30}$$

Equation (28) can be written as:

$$\begin{cases} \sigma_\alpha \dot\sigma_\alpha = \left(\left(\overbrace{k_1 \dot i^*_{inv\alpha}}^{term1} - \overbrace{\frac{k_1 v_{inv\alpha}}{L_{inv}} + \frac{k_1 v_{L\alpha}}{L_{inv}}}^{term2} + \overbrace{k_{i1} i^*_{inv\alpha}}^{Term3} - \overbrace{\frac{R_{inv}}{L_{inv}} i_{inv\alpha} - k_{i1} i_{inv\alpha}}^{term4} \right) \right) \left(\overbrace{k_1 e_\alpha(t) + k_{i1}}^{term5} \overbrace{\int_0^t e_\alpha(i_{inv\alpha}) dt}^{term6} \right) < 0 \\[4em] \sigma_\beta \dot\sigma_\beta = \left(\left(\overbrace{k_1 \dot i^*_{inv\beta}}^{term1} - \overbrace{\frac{k_1 v_{inv\beta}}{L_{inv}} + \frac{k_1 v_{L\beta}}{L_{inv}}}^{term2} + \overbrace{k_{i1} i^*_{inv\beta}}^{Term3} - \overbrace{\frac{R_{inv}}{L_{inv}} i_{inv\beta} - k_{i1} i_{inv\beta}}^{term4} \right) \right) \left(\overbrace{k_1 e_\beta(t) + k_{i1}}^{term5} \overbrace{\int_0^t e_\beta(i_{inv\beta}) dt}^{term6} \right) < 0 \end{cases} \tag{31}$$

To determine the sign of (31), one determines first the sign of each term independently. Considering the first term ($k_1 \dot i^*_{inv\alpha}$), which represents the derivative of the reference inverter current, is equal to zero and regarding, that DG ensures the regulation of the voltage at PCC, one assumes that the PCC voltage and the voltage at the output of the VSC are equal, which leads that the second term ($-\frac{k_1 v_{inv\alpha}}{L_{inv}} + \frac{k_1 v_{L\alpha}}{L_{inv}}$) being equal to zero, and $k_{i1} i^*_{inv\alpha} \ll -\frac{R_{inv}}{L_{inv}} i_{inv\alpha} - k_{i1} i_{inv\alpha}$. As known the integral is always positive, so the sign of the sixth term ($k_{i1} \int_0^t e_\alpha(i_{inv\alpha}) dt$) is positive. The error between the inverter current and its reference is smaller, so, the fifth term ($k_1 e_\alpha(t)$) is smaller than the sixth term. Based on this analysis, one concludes that the sign of $\sigma_\alpha \dot\sigma_\alpha$ is always negative, which implies that the system is asymptotically stable.

4. Results and Discussion

The performance of the developed control strategies for the VSC and DC-DC buck-boost converter, was tested through a simulation using MATLAB/Simulink and in real-time using a hardware prototype. In Figure 6 the hardware prototype of a standalone microgrid is presented. It consists of: (1) ABB drive, (2) synchronizer of DG, (3) induction motor, (4) synchronous generator, (5) SPVA simulator, (6) loads, (7) Protections cards, dSPACE and SEMIKRON power converter, (8) sensors, (9) step down transformer, (10) RC passive filter, and (11) lead acid battery pack.

Figure 6. Hardware prototype of a standalone microgrid.

In Figure 7 the dynamic performances of the terminal voltage (v_{DG}) and current i_{DG} of synchronous generator, load current i_L, inverter current (i_{inv}), DC-link voltage (V_{dc}) and its reference (v^*_{dc}), battery current (v_{DG}) and its reference (v_{DG}), and output SPVA current (I_{PV}) are shown.

The zoomed waveforms presented in Figure 6 are shown in Figure 8. It is observed that the AC voltage at the PCC and the frequency are regulated at their rated values, which confirms that the governor and the automatic voltage regulator of the DG are working well. One can see clearly in the zoomed waveforms presented in Figure 8a–c, that the DG currents are balanced and sinusoidal, which confirms that the single-phase d-q control-based sliding mode controller performs well in the presence of balanced nonlinear loads. One observes that the DC-link voltage and battery current follow their references during solar irradiation changes, which confirms the robustness of the developed control strategy-based sliding mode control with boundary layer. One observes from the obtained results shown in Figure 8a–c that during transitions when solar irradiation changes the battery current follows its reference, which represents the output of the DC-link voltage regulator without saturation and with zero steady state error. The output SPVA current, which represents the maximum of SPVA current and DC-link voltage varies with variation of solar irradiation as shown in Figure 3, which confirms that it is possible to achieve MPPT without using any MPPT method.

In Figures 9 and 10, the waveforms and the zoomed of terminal voltage (v_{DG}) and current i_{DG} of synchronous generator, load current i_L, inverter current (i_{inv}), DC-link voltage (V_{dc}) and its reference (v_{dc}^*), battery current (v_{DG}) and its reference (v_{DG}), and output SPVA current (I_{PV}), are presented. To test the performance of the developed control strategies, the system is subjected to severe conditions, such as the presence of balanced and unbalanced nonlinear RL and RC load types as well as during solar irradiation changes. It is observed that in the presence of different loads, the DG current is balanced and sinusoidal.

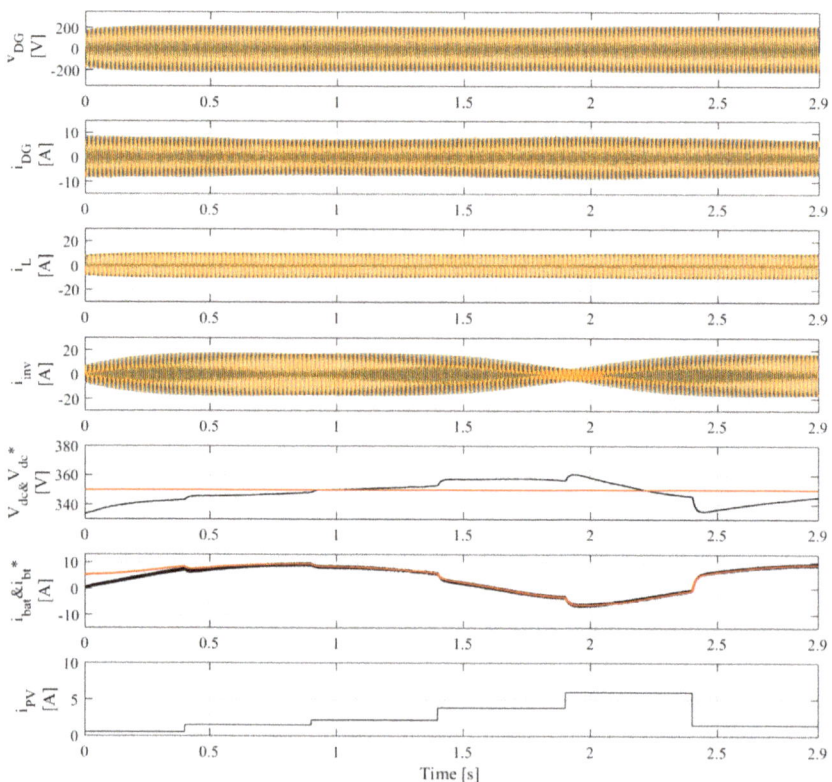

Figure 7. Dynamic performance under presence of balanced RL nonlinear load and solar irradiation change.

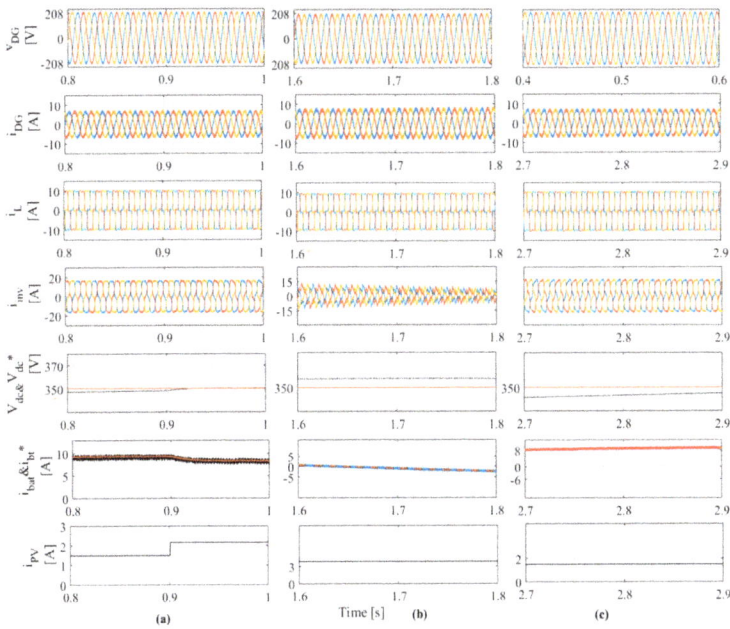

Figure 8. Zoom of waveforms of Figure 6 between (**a**) t = 0.8 s to t = 1 s, (**b**) t = 1.6 s to t = 1.8 s and (**c**) between t = 2.7 s to t = 2.9 s.

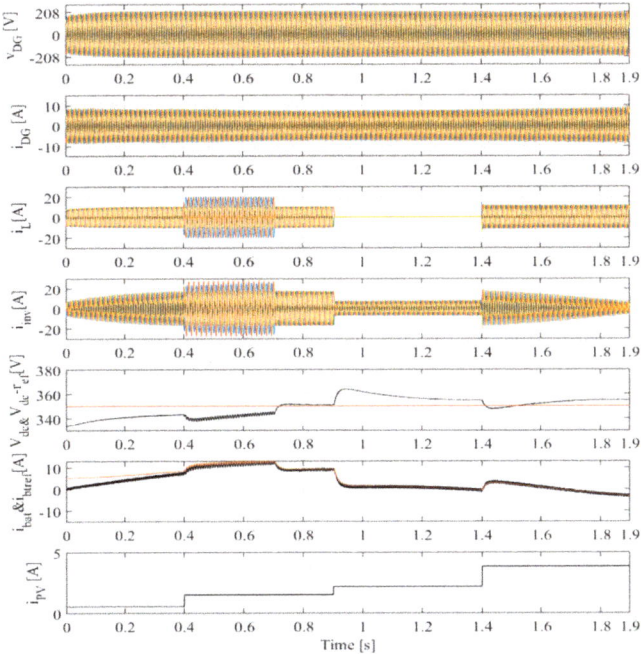

Figure 9. Dynamic performance under presence of balanced and unbalanced nonlinear load (diode bridge +RL and RC loads) and solar irradiation change.

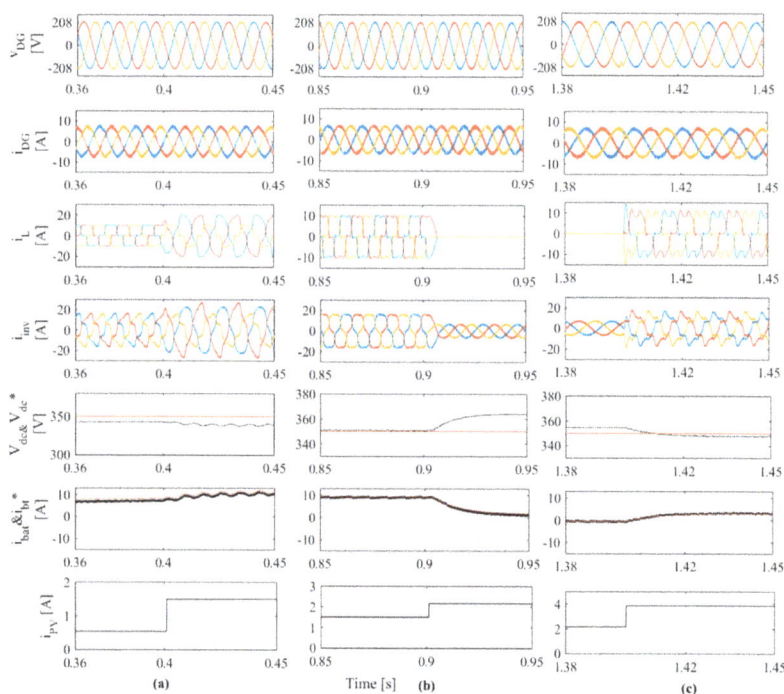

Figure 10. Zoom of the waveforms of Figure 6 between (**a**) t = 0.8s to t = 1 s, (**b**) t = 1.6 s to t = 1.8 s and (**c**) between t = 2.7 s to t = 2.9 s.

In addition, DC-DC buck boost can extract the maximum of power from the SPVA and balance the power in the system by charging and discharging the battery, which confirms the robustness of the proposed control strategies based on sliding mode control for the standalone microgrid system. In Figure 11 the harmonics spectrum of the PCC voltage, DG current, and load currents when RL and RC nonlinear load, are presented. It is observed that in the both cases the total harmonic distortion (THD) of the load current is high (THD = 24% with RL nonlinear and 27% with RC nonlinear load), and the THD of the source current, which represents in the standalone microgrid shown in Figure 1, the output current of the DG is less than 5% (THD = 4.07% with RL nonlinear and 4.28% with RC nonlinear load). This respects the norms of IEEE standard 519-1992. One sees clearly that the THD of the PCC voltage is less than to 5% in the presence of the both nonlinear loads, which confirms the robustness of the proposed Single-phased d-q theory-based on SMC for power quality improvement.

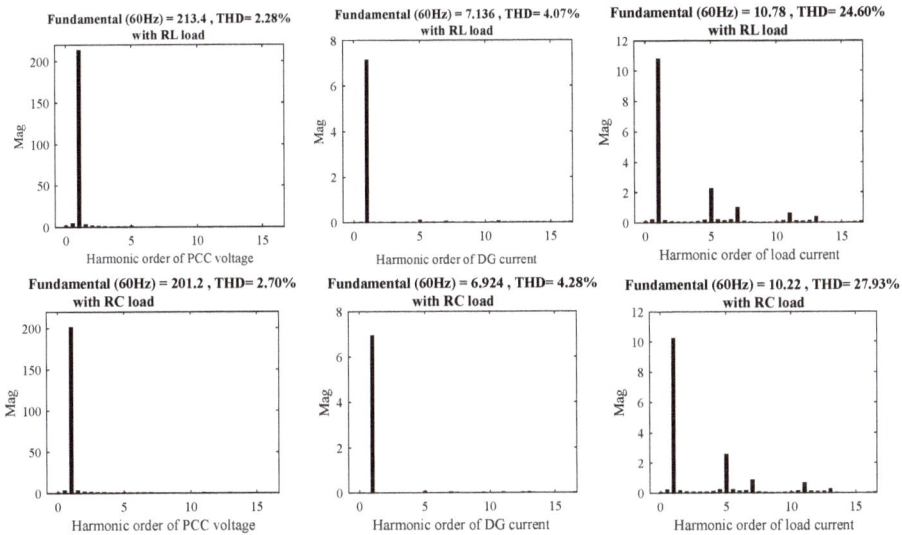

Figure 11. Harmonics spectrum of the PCC voltage, DG current and load current with RL and RC nonlinear loads.

In Figure 12a,b, the steady-state performance of the DG is presented. One observes that the terminal voltage before step down transformer (v_{DGa1}) and at the secondary of transformer v_{DGa2} are regulated constant and sinusoidal. Seeing that DG operates at fixed speed, excitation current (i_{exc}) is constant, one can observe that the system frequency is constant, and the stator current are sinusoidal.

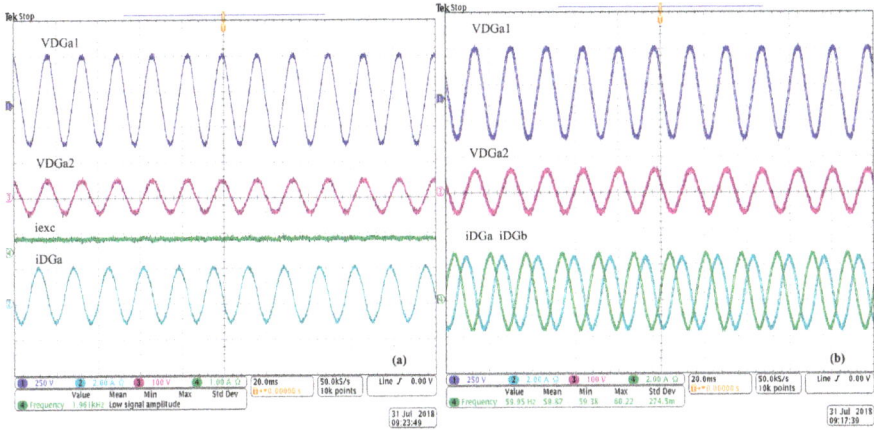

Figure 12. Steady-state performance in the presence of linear loads.

To test the performance of the developed control strategies-based sliding mode control, the system is subjected to load variations as presented in Figure 13a–c. It is observed in Figure 13a that at t = 0.2 ms the system is subjected to a sudden variation of nonlinear load type RL. One observes that VSC operates as a shunt active filter, and it compensates the harmonics and balances the DG current. One observes that the DC-link voltage is well regulated. In Figure 13b, the system is subjected to sudden switching off the nonlinear load type RC at t = 0.2 ms. It is observed that the VSC acts as a shunt filter and compensates the harmonics as well as balances the DG current. One sees that DC-link voltage is increased when the load is switched off, because the battery starts charging. In Figure 13c, the system is subjected to sudden switching on and off at t = 0.1 ms and at t = 1 s, and one observes that the DC-link voltage is well regulated and the VSC acts as a shunt active filter and injects the power into the PCC. One sees clearly in Figure 13a–c, that the both developed control strategies perform well in the presence of severe conditions.

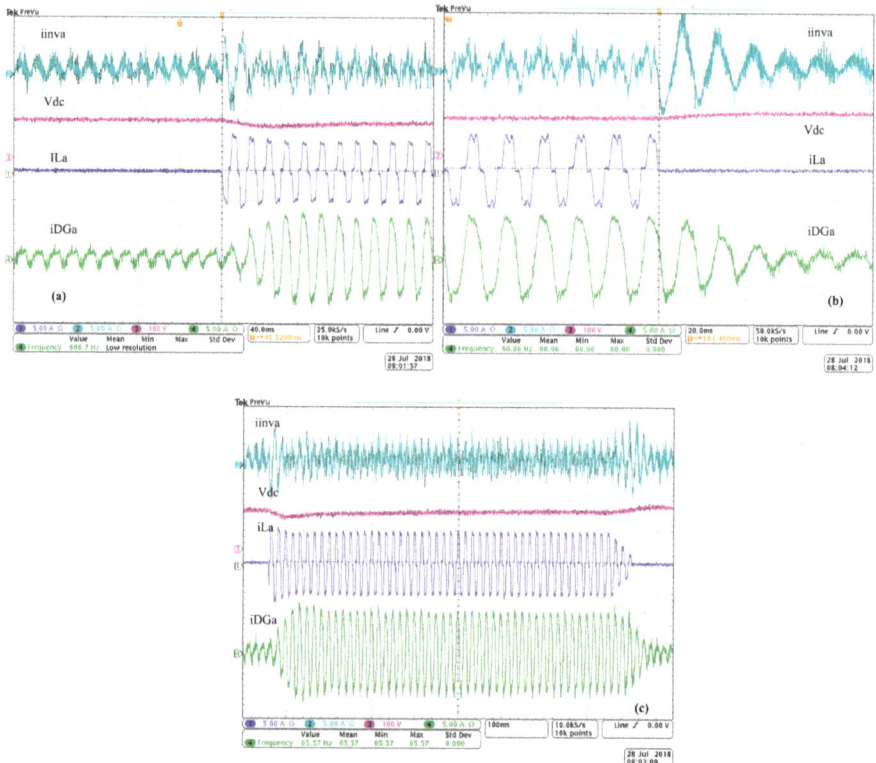

Figure 13. Dynamic performance under sudden variation of: (**a**) balanced RL nonlinear load, (**b**) balanced RC nonlinear load, and (**c**) under switching on and off of RL nonlinear load.

The performance of DC-link voltage (V_{dc}), battery current (i_{bat}), load current (i_{La}) and inverter current (i_{inva}), are presented in Figure 14a,b. It is observed that the DC-link voltage is regulated at their rated value when the system is subjected to load variations at t = 0.4 s and t = 0.8 s in Figure 14a, and at t = 0.3 s, and t = 0.8 s in Figure 14b. One observes that the battery current varies with load variation, balancing the power in the system. This leads that the system performing well during all conditions, which confirms the robustness of the developed control strategy-based sliding mode control.

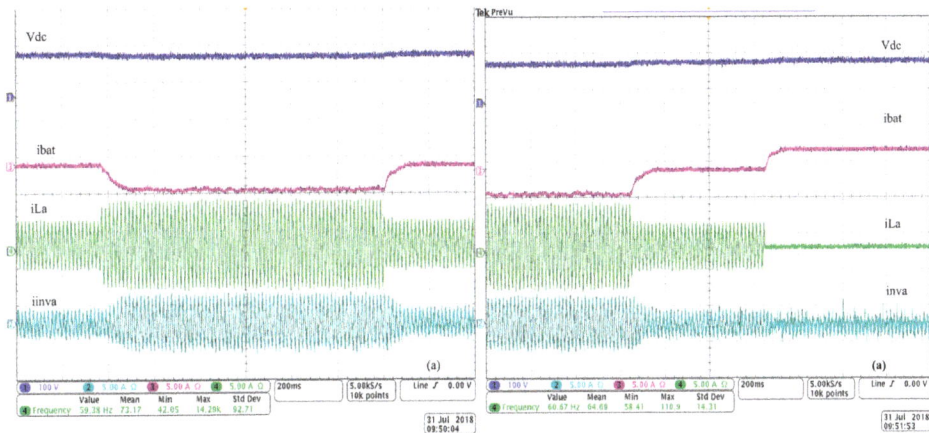

Figure 14. Dynamic performance under fixed solar irradiation and load change. (**a**) sudden increase and decrease of load; and (**b**) sudden decrease and disconnect of load.

5. Conclusions

In this paper, a real-time implementation of robust control strategies based on sliding mode control for a standalone microgrid has been presented. Single-phase d-q theory based on sliding mode controller for fast inner control loop for three-phase voltage source converter was studied and analyzed. Furthermore, modeling and stability analyses are given in detail. Sliding mode control with boundary layer is developed for DC-DC buck boost. It has been proven that the developed control strategies perform well during severe conditions, such as load variation and solar irradiation changes. The obtained simulation, as well as, experimental results show satisfactory performance.

Author Contributions: This paper was a collaborative effort between the authors. The authors under supervision of A.C. contributed collectively in all steps of this research work including validation in real-time.

Funding: This research received no external funding.

Acknowledgments: Validation in real time is realized at École de technologie supérieure in laboratory of Power Electronics and Industrial Control Research Group (GREPCI), at Montreal, Quebec, Canada.

Conflicts of Interest: The authors declare no conflict of interest.

References

1. Kaldellis, J.K. *Stand-Alone and Hybrid Wind Energy Systems: Technology, Energy Storage and Applications*, 1st ed.; Woodhead Publishing: Cambridge, MA, USA, 2010; pp. 7–10.
2. Sreekumar, P.; Khadkikar, V. Adaptive power management strategy for effective volt–ampere utilization of a photovoltaic generation unit in standalone microgrids. *IEEE Trans. Ind. Appl.* **2018**, *54*, 1784–1792. [CrossRef]
3. Rezkallah, M.; Chandra, A.; Tremblay, M.; Ibrahim, H. Experimental Implementation of an APC with Enhanced MPPT for Standalone Solar Photovoltaic based Water Pumping Station. *IEEE Trans. Sustain. Energy* **2018**. [CrossRef]
4. Shi, R.; Zhang, X. VSG-Based dynamic frequency support control for autonomous PV–diesel microgrids. *Energies* **2018**, *11*, 1814. [CrossRef]
5. Rezkallah, M.; Singh, S.; Chandra, A.; Saad, M.; Singh, B.; Tremblay, M.; Geng, H. Real-time hardware testing, control and performance analysis of hybrid cost-effective wind-PV-diesel standalone power generation system. In Proceedings of the 2017 IEEE Industry Applications Society Annual Meeting, Cincinnati, OH, USA, 1–5 October 2017; pp. 1–8.
6. Gan, L.K.; Shek, J.K.; Mueller, M.A. Analysis of Tower Shadow Effects on Battery Lifetime in Standalone Hybrid Wind-Diesel-Battery Systems. *IEEE Trans. Ind. Electron.* **2017**, *64*, 6234–6244. [CrossRef]

7. Nguyen, N.H.; Nguyen-Duc, H.; Nakanishi, Y. Optimal sizing of energy storage devices in wind-diesel systems considering load growth uncertainty. *IEEE Trans. Ind. Appl.* **2018**, *54*, 1983–1991. [CrossRef]
8. Rezkallah, M.; Chandra, A.; Rousse, D.R.; Ibrahim, H.; Ilinca, A.; Ramdenee, D. Control of small-scale wind/diesel/battery hybrid standalone power generation system based on fixed speed generators for remote areas. In Proceedings of the IEEE, IECON 2016, Florence, Italy, 23–26 October 2016; pp. 4060–4065.
9. Kamal, E.; Aitouche, A.; Oueidat, M. Fuzzy fault-tolerant control of wind-diesel hybrid systems subject to sensor faults. *IEEE Trans. Sustain. Energy* **2013**, *4*, 857–866. [CrossRef]
10. Adefarati, T.; Bansal, R.C.; Justo, J.J. Techno-economic analysis of a PV–wind–battery–diesel standalone power system in a remote area. *J. Eng.* **2017**, *13*, 740–744. [CrossRef]
11. Rezkallah, M.; Hamadi, A.; Chandra, A.; Singh, B. Design and implementation of active power control with improved P&O method for wind-PV-battery-based standalone generation system. *IEEE Trans. Ind. Electron.* **2018**, *65*, 5590–5600.
12. Rezkallah, M.; Chandra, A.; Singh, B.; Singh, S. Microgrid: Configurations, Control and Applications. *IEEE Trans. Smart Grid* **2017**. [CrossRef]
13. Wies, R.W.; Johnson, R.A.; Agrawal, A.N.; Chubb, T.J. Simulink model for economic analysis and environmental impacts of a PV with diesel-battery system for remote villages. *IEEE Trans. Power Syst.* **2005**, *20*, 692–700. [CrossRef]
14. Nejabatkhah, F.; Li, Y.W.; Nassif, A.B.; Kang, T. Optimal design and operation of a remote hybrid microgrid. *CPSS Trans. Power Electron. Appl.* **2018**, *3*, 3–13. [CrossRef]
15. Mokhtar, M.; Marei, M.I.; El-Sattar, A.A. An adaptive droop control scheme for DC microgrids integrating sliding mode voltage and current controlled boost converters. *IEEE Trans. Smart Grid* **2017**. [CrossRef]
16. Yazici, İ.; Yaylaci, E.K. Fast and robust voltage control of DC–DC boost converter by using fast terminal sliding mode controller. *IET Power Electron.* **2016**, *9*, 120–125. [CrossRef]
17. Chincholkar, S.H.; Jiang, W.; Chan, C.Y. An improved PWM-based sliding-mode controller for a DC-DC cascade boost converter. *IEEE Trans. Circ. Syst. II Express Br.* **2017**. [CrossRef]
18. Singh, B.; Chandra, A.; Al-Haddad, K. *Power Quality: Problems and Mitigation Techniques*; John Wiley & Sons: Hoboken, NJ, USA, 2014; pp. 2–10.
19. Dong, H.; Yuan, S.; Han, Z.; Ding, X.; Ma, S.; Hana, X. A Comprehensive strategy for power quality improvement of multi-inverter-based microgrid with mixed loads. *IEEE Access* **2018**, *6*, 30903–30916. [CrossRef]
20. Ko, H.S.; Jang, M.S.; Ryu, K.S.; Kim, D.J.; Kim, B.K. Supervisory Power Quality Control Scheme for a Grid-Off Microgrid. *IEEE Trans. Sustain. Energy* **2018**, *9*, 1003–1010. [CrossRef]
21. Kesler, M.; Ozdemir, E. Synchronous-reference-frame-based control method for UPQC under unbalanced and distorted load conditions. *IEEE Trans. Ind. Electron.* **2011**, *58*, 3967–3975. [CrossRef]
22. García, P.; García, C.A.; Fernández, L.M.; Llorens, F.; Jurado, F. ANFIS-based control of a grid-connected hybrid system integrating renewable energies, hydrogen and batteries. *IEEE Trans. Ind. Inform.* **2014**, *10*, 1107–1117. [CrossRef]
23. Seifi, K.; Moallem, M. An adaptive PR controller for synchronizing grid-connected inverters. *IEEE Trans. Ind. Electron.* **2018**. [CrossRef]
24. Wang, T.; Fei, J. Adaptive neural control of active power filter using fuzzy sliding mode controller. *IEEE Access* **2016**, *4*, 6816–6822. [CrossRef]
25. Fei, J.; Chu, Y.; Hou, S. A backstepping neural global sliding mode control using fuzzy approximator for three-phase active power filter. *IEEE Access* **2017**, *5*, 16021–16032. [CrossRef]
26. Cao, D.; Fei, J. Adaptive fractional fuzzy sliding mode control for three-phase active power filter. *IEEE Access* **2016**, *4*, 6645–6651. [CrossRef]
27. Guzman, R.; de Vicuña, L.G.; Morales, J.; Castilla, M.; Miret, J. Model-based control for a three-phase shunt active power filter. *IEEE Trans. Ind. Electron.* **2016**, *63*, 3998–4007. [CrossRef]
28. Tsai, J.F.; Chen, Y.P. Sliding mode control and stability analysis of buck DC-DC converter. *Int. J. Electron.* **2007**, *94*, 209–222. [CrossRef]
29. Reitz, M.A.; Wang, X. Robust Sliding Mode Control of Buck-Boost DC-DC Converters. In Proceedings of the ASME 2016 Dynamic Systems and Control Conference, MN, USA, 12–14 October 2016.
30. Guldemir, H. Modeling and sliding mode control of DC-DC buck-boost converter. In Proceedings of the 6th International Advanced Technologies Symposium (IATS'11), Elazığ, Turkey, 16–18 May 2011; pp. 475–480.

31. Slotine, J.J.; Li, W. *Applied nonlinear Control*; Prentice Hall: Englewood Cliffs, NJ, USA, 1991; pp. 105–116.

32. Gavagsaz-Ghoachani, R.; Phattanasak, M.; Martin, J.P.; Pierfederici, S.; Nahid-Mobarakeh, B.; Riedinger, P. A lyapunov function for switching command of a DC–DC power converter with an LC input filter. *IEEE Trans. Ind. Appl.* **2017**, *53*, 5041–5050. [CrossRef]

33. Rezkallah, M.; Sharma, S.K.; Chandra, A.; Singh, B.; Rousse, D.R. Lyapunov function and sliding mode control approach for the solar-PV grid interface system. *IEEE Trans. Ind. Electron.* **2017**, *64*, 785–795. [CrossRef]

34. Rezkallah, M.; Hamadi, A.; Chandra, A.; Singh, B. Real-time HIL implementation of sliding mode control for standalone system based on PV array without using dumpload. *IEEE Trans. Sustain. Energy.* **2015**, *6*, 1389–1398. [CrossRef]

35. Wai, R.J.; Shih, L.C. Design of voltage tracking control for DC–DC boost converter via total sliding-mode technique. *IEEE Trans. Ind. Electron.* **2011**, *58*, 2502–2511. [CrossRef]

36. Saghafinia, A.; Ping, H.W.; Uddin, M.N. Fuzzy sliding mode control based on boundary layer theory for chattering-free and robust induction motor drive. *Int. J. Adv. Manuf. Technol.* **2014**, *71*, 57–68. [CrossRef]

37. Wang, W.J.; Chen, J.Y. A new sliding mode position controller with adaptive load torque estimator for an induction motor. *IEEE Trans. Energy Convers.* **1999**, *14*, 413–418. [CrossRef]

38. Boum, A.T.; Djidjio Keubeng, G.B.; Bitjoka, L. Sliding mode control of a three-phase parallel active filter based on a two-level voltage converter. *Syst. Sci. Control Eng.* **2017**, *5*, 535–543. [CrossRef]

energies

MDPI

Article

An Improved Multi-Timescale Coordinated Control Strategy for Stand-Alone Microgrid with Hybrid Energy Storage System

Jingfeng Chen [1,2], Ping Yang [1,2,3], Jiajun Peng [1,*], Yuqi Huang [1], Yaosheng Chen [1] and Zhiji Zeng [1]

[1] School of Electric Power, South China University of Technology, Guangzhou 510640, China; wanwan0124@gmail.com (J.C.); eppyang@scut.edu.cn (P.Y.); epcaceros@mail.scut.edu.cn (Y.H.); 201621012049@mail.scut.edu.cn (Y.C.); zeng.zhiji@mail.scut.edu.cn (Z.Z.)
[2] Guangdong Key Laboratory of Clean Energy Technology, Guangzhou 511458, China
[3] National-Local Joint Engineering Laboratory for Wind Power Control and Integration Technology, South China University of Technology, Guangzhou 511458, China
* Correspondence: peng.jiajun@mail.scut.edu.cn; Tel.: +86-135-7042-5742

Received: 29 July 2018; Accepted: 15 August 2018; Published: 17 August 2018

Abstract: A scientific and effective coordinated control strategy is crucial to the safe and economic operation of a microgrid (MG). With the continuous improvement of the renewable energy source (RES) penetration rate in MG, the randomness and intermittency of its output lead to the increasing regulation pressure of the conventional controllable units, the increase of the operating risk of MG and the difficulty in improving the operational economy. To solve the mentioned problems and take advantage of hybrid energy storage system (HESS), this study proposes a multi-time scale coordinated control scheme of "day-ahead optimization (DAO) + intraday rolling (IDR) + quasi-real-time correction (QRTC) + real-time coordinated control (RTCC)." Considering the shortcomings of existing low prediction accuracy of distributed RES and loads, the soft constraints such as unit commitment scheduling errors and load switching scheduling errors are introduced in the intraday rolling model, allowing the correction of day-ahead unit commitment and load switching schedule. In the quasi-real-time coordinated control, an integrated criterion is introduced to decide the adjustment priority of the distributed generations. In the real-time coordinated control, the HESS adopts an improved first order low pass filtering algorithm to adaptively compensate the second-level unbalanced power. Compared with the traditional coordinated control strategy, the proposed improved model has the advantages of good robustness and fast solving speed and provides some guidance for the intelligent solution for stable and economic operation of stand-alone MG with HESS.

Keywords: hybrid energy storage; stand-alone microgrid; multi-time scale; coordinated control

1. Introduction

In recent years, distributed generation (DG) technology has rapidly developed due to its advantage of efficiently consuming energy locally. To fully take advantage of DG and improve the safety and reliability of power supply, the MG is proposed as an effective scheme to improve the DG penetration rate of the distribution network.

MG consists of distributed generations (DGs), energy storage system (ESS), power load, monitoring, protection and automation devices. It can be viewed as a small power supply system that can realize the internal power balance. MG can operate in many modes. It can connect to the distribution network through point of common coupling (PCC), operating as an "equivalent controllable load." Or it can be disconnected from the distribution network operating in stand-alone mode and provides power for the interior load [1,2]. For occasions where it is not feasible to establish distribution network, MG can only operate independently and autonomously [3–6]. For the stand-alone MG, the current mainstream operation control strategies are master-slave strategy and peer-to-peer strategy [7]. In master-slave strategy, one of the DGs (diesel generators, PV system, ESS, etc.) uses V/f control method, producing voltage and frequency references for other DGs and other DGs use P/Q control method. While in peer-to-peer strategy, each DG that participates in V/f regulation and control plays an equivalent role. In this strategy, DG controllers usually use droop control, automatically dispatch the output power.

A scientific and effective coordinated control strategy is crucial to the safe and economic operation of MG. Current researches on coordinated control of stand-alone MGs can be categorized into two groups: coordinated control strategies based on fixed logic criteria and strategies based on optimization theory. Reference [8] targets at stand-alone MG that contains wind turbine (WT), photovoltaic (PV) system, diesel generators (DSGs) and ESS and summarizes several commonly used coordinated control strategies based on logic criteria, including power smoothing strategy, minimum running time strategy for DSGs, soft cycle charging strategy, hard cycle charging strategy and so forth. Since the fixed logic criteria based control strategy is based on pre-analysis and operational experience and does not change with the load or RES, it is easy to design and has high decision-making speed. Therefore, most of the stand-alone MGs adopt this kind of strategy.

Due to the volatility and randomness of RES, the fixed logic criteria based coordinated control strategy cannot guarantee optimal economic operation of MG. Therefore, the strategy based on optimization theory has been extensively studied by scholars. An improved simplified warm optimization method for day-ahead operation optimization model is proposed in [9]. The day-ahead model includes fuel cost, battery operation cost and power transmission cost. In [8], an economic dispatch model considering battery lifetime for MG is proposed. A scenario-based robust energy management method accounting for the worst-case amount of renewable generation (RG) and load is developed in [10]. Reference [11] presented an optimal management of battery energy storage in a PV-based commercial building to increase its resilience as it minimizes its operational cost. The Conditional Value at Risk (CVaR) was used to account for the uncertainties in both the day-ahead electricity price and the PV power generation. To deal with the prediction uncertainties of RE and loads and take advantage of the time-of-use electricity price, reference [12] developed an interval-based optimization model for maximum profits.

The above papers are based on day-ahead prediction results, focusing on the research of day-ahead scheduling optimization model to reduce the risk brought by the uncertainty of RES. However, compared to centralized RES, the prediction accuracy of distributed RES is poor, so the schedule result may not be directly applied in MG. Reference [13] shows that the shorter the prediction horizon, the higher the prediction accuracy. So, the using of control strategy with smaller timescale can correct the residual errors produced in larger timescale. Therefore, ultra-short-term prediction with smaller timescale is applied to the IDR optimization in MG and multi-timescale coordinated control strategies have emerged. Reference [14] developed an energy management framework for MG including multi-timescale demand response. The timescales are days, hours and minutes. The hour-ahead scheduling model is based on model prediction control, with prediction data generated by ultra-short-term prediction. Reference [15] proposes a coordinated control strategy in day-ahead and intraday aspects, considering battery lifetime degradation cost when optimizes operation cost. In [14,15], the startup and shutdown schedule of controllable generation units cannot be modified. When the day-ahead prediction error is high, it is possible that the generation power cannot balance load power. [16] proposes a two-timescales robust optimization method by scheduling energy storage and the direct load control. Reference [17] proposed an energy management model for MG based on day-ahead and real-time timescale. Day-ahead power prediction is used in the decision-making of next-day economic operation of MG. A fuzzy control based supervisory control strategy is adopted in real-time timescale to reduce the tie line power deviation. In [18], a strategy for obtaining optimal scheduling of multiple microgrid systems with power sharing through coordination among microgrids that have no cost function of generation units is proposed.

However, these papers have the following shortcomings:

(1)   HESS is not included as an element of coordinated control; thus, applicability is limited.
(2)   The SOC balance of HESS in daily dispatch period has not received enough attention.
(3)   The multi-timescale framework is relatively simple, with large gap between different timescales.

At present, energy-type ESS (ETESS) such as batteries are usually used as the main energy storage device in MG. ETESS has the characteristics of high energy density, low power density, short cycle life and slow power response, thus cannot economically smooth high frequency disturbance in MG. Power-type ESS (PTESS) represented by supercapacitor (SC) has the characteristics of high power density, low energy density, long cycle life and fast power response. Therefore, a better solution is to combine ETESS and PTESS, forming HESS. According to [5], it is proved theoretically that HESS can make full use of the complementary advantages of battery and SC, improve the output performance of ESS and prolong the service life of battery, reducing the life cycle cost of ESS. By now, many research has been conducted on the optimal configuration and control methods for HESS [19–21] but only a few research on energy management and coordinated control for MG wit HESS. Reference [22] proposed a two-layer energy management framework for HESS composed of battery and SC and incorporated the battery cycle charge and discharge life loss into the model. However, this study is only for MG with RES, HESS and load, limited for the scenario where other controllable power sources (for example, diesel generators and micro gas turbine) exist in MG. Reference [23] proposed a two-layer energy scheduling framework for MG: hour-ahead scheduling and real-time scheduling. In hour-ahead scheduling, a deterministic optimization model is formulated to minimize the operation cost of microgrids and to guarantee the operation safety. The real-time scheduling is conducted every minute and the scheduling period is still too long compared to the SC with a charge and discharge cycle of only tens of seconds.

In this paper, an improved multi-time scale coordinated control strategy is proposed for stand-alone MG with HESS. The strategy consists of day-ahead optimization model, model predictive control based intraday rolling optimization model, comprehensive criteria based quasi-real-time correction model and real-time coordinated control model. Compared with the traditional coordinated control strategy, the proposed improved model has the advantages of good robustness and fast solving speed and provides some guidance for the intelligent solution for stable and economic operation of stand-alone MG with HESS.

The remainder of this paper is organized as follows. Section 2 describes the typical topology of stand-alone microgrid with hybrid energy storage system. Section 3 presents the multi-time scale coordinated control proposed in this paper. Section 4 demonstrates and analyzes the simulation results of a case study. Summary, conclusions and outlook are given in Section 5.

2. Typical Topology of Stand-Alone Microgrid with Hybrid Energy Storage System

Figure 1 is a typical topology of stand-alone MG with HESS. In this MG, wind power generation units, PV units, diesel generation units, HESS and load are connected to the AC bus through multiple feeders. Feeders are usually equipped with circuit breaker and the entire grid has a radial structure.

The hierarchical control architecture is adopted to allocate the secondary system of stand-alone MG and each layer has a corresponding control unit to execute the control strategy:

(1)   The first layer is the optimal schedule layer and the corresponding control unit is EMS (energy management system). The DAO and IDR algorithm in EMS is used to realize coordinated control strategy of a large time scale and provide scheduling curve for MGCC (microgrid central controller).

(2) The second layer is the MG control layer and the corresponding control unit is MGCC. The QRTC and RTCC in MGCC based on logical judgment is used to correct the deviation between the actual operating state and the ideal state of the MG and issue control commands to the device controller.

(3) The third layer is the local control layer and the corresponding control unit is device controllers. Control commands from the MG real-time control layer are executed by DGs/load/HESS.

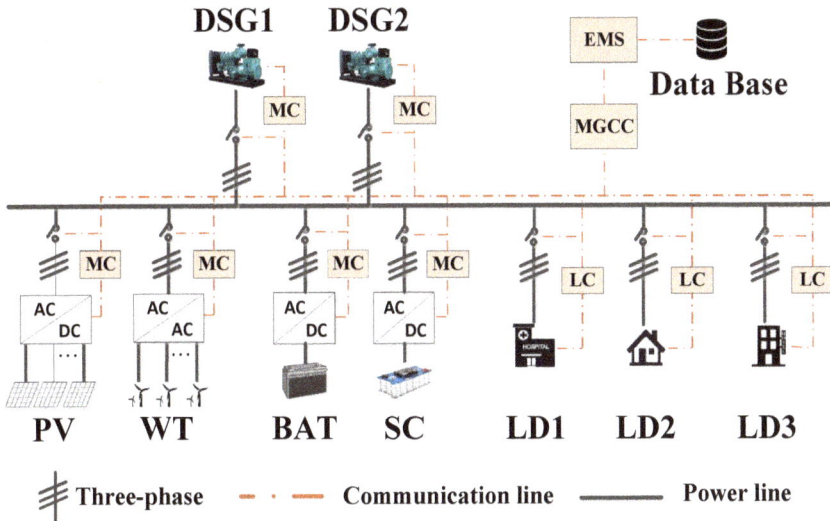

Figure 1. A typical topology of MG with HESS.

To coordinate various DGs in MG, it is necessary to model the operating characteristics and boundaries of DGs, including the wind power system operation model, PV power system operation model, diesel power generation system operation model, the charge/discharge model of HESS and energy storage life model and so forth. The wind power and PV power system operation model re implemented by [18], the diesel model is implemented by [15], the HESS model is implemented by [24,25] and the life model is implemented by [26]. These models will not be further discussed here.

3. Multi-time Scale Coordinated Control

3.1. Multi-time Scale Coordinated Control Framework

The framework of multi-time scale coordinated control is shown in Figure 2. This framework is comprised of day-ahead optimization model, intraday rolling scheduling model, quasi-real-time coordinated control model and real-time control model, which are described as follows:

(1) Day-ahead optimization model: Based on the short-term load forecasting results, DGs' hourly output scheduling curve (except PTESS) and load switching scheme are determined by DAO Day-ahead optimization model in the EMS. The execution cycle of this model is 1 day.

(2) Intraday rolling model: Based on the ultra short-term power forecasting results, the DAO results are continuously corrected by the time-limited rolling optimization scheduling. The time scale of the rolling time window is 4h (same as the ultra short-term prediction). The model has an execution cycle of 15 min.

(3) Quasi-real-time coordinated control model: The deviation between the constantly updated intraday rolling scheduling plan and the actual operating conditions of MG is calculated by the real-time collected data and will be corrected quickly. An integrated criterion is introduced to

decide the adjustment priority of the distributed generations. The execution cycle of this model is 1 min.

(4) Real-time coordinated control model: Under the condition of ensuring frequency and voltage stability, real-time control commands are determined to follow the command from quasi-real-time coordinated control as much as possible in second time scale and real-time smooth control strategy for HESS based on logic judgment is used to smooth unbalanced power of second time scale. The execution cycle of this model is 5 s.

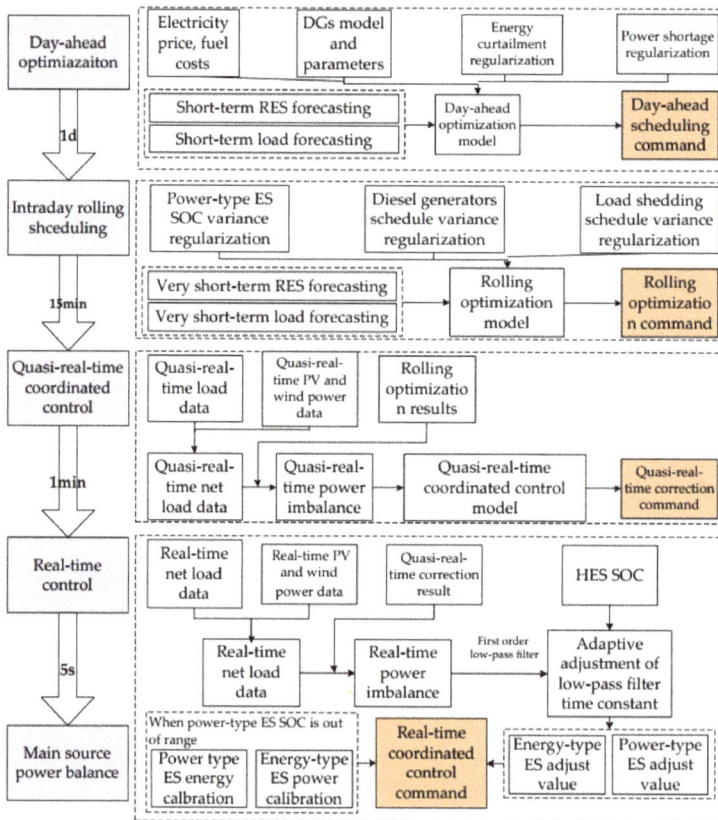

Figure 2. The framework of multi-time scale coordinated control.

3.2. Day-Ahead Optimization Model

3.2.1. Decision Variables

This paper assume that stand-alone MG includes WT system, PV system, ETESS, PTESS and DSGs on the source side and important load, secondary load and interruptible load of participation in demand response on the load side. The optimization variables of the MG's day-ahead optimization scheduling model include the set *P* of output scheduling curves of controllable DGs and the set *u* of DG unit commitment and load switching schedule:

$$P = \left\{ P_{de,i}, P_{ba,j}, P_{pv,l}, P_{wt,m} \right\}; u = \left\{ u_{de,i}, u_{ba,j}, u_{sdload,k}, u_{itload,p} \right\}$$

$$
\begin{cases}
P_{de} = \left\{ P_{de,1}, P_{de,2}, \cdots, P_{de,i}, \cdots, P_{de,n_{de}} \right\} \\
P_{ba} = \left\{ P_{ba,1}, P_{ba,2}, \cdots, P_{ba,j}, \cdots, P_{ba,n_{ba}} \right\} \\
P_{pv} = \left\{ P_{pv,1}, P_{pv,2}, \cdots, P_{pv,l}, \cdots, P_{pv,n_{pv}} \right\} \\
P_{wt} = \left\{ P_{wt,1}, P_{wt,2}, \cdots, P_{wt,m}, \cdots, P_{wt,n_{wt}} \right\} \\
u_{de} = \left\{ u_{de,1}, u_{de,2} \cdots u_{de,k}, \cdots, u_{de,n_{de}} \right\} \\
u_{ba} = \left\{ u_{ba,1}, u_{ba,2}, \cdots, u_{ba,j}, \cdots, u_{ba,n_{ba}} \right\} \\
u_{sdload} = \left\{ u_{sdload,1}, u_{sdload,2}, \cdots, u_{sdload,k}, \cdots, u_{sdload,n_{sdload}} \right\} \\
u_{itload} = \left\{ u_{itload,1}, u_{itload,2}, \cdots, u_{itload,p}, \cdots, u_{itload,n_{itload}} \right\}
\end{cases}
\tag{1}
$$

where $n_{de}, n_{ba}, n_{pv}, n_{wt}, n_{sdload}, n_{itload}$ are the number of diesel generators, ETESS, PV system, wind power system, secondary load and interruptible load. $P_{de}, P_{ba}, P_{pv}, P_{wt}$ are the output schedule matrix of diesel generators, ETESS, PV system and wind power system. u_{de}, u_{ba} are the operation state matrix of diesel generators and ETESS, where 0 means stopped and 1 means running. u_{sdload}, u_{ctload} are load switching schedule matrix of secondary load and interruptible load, where 0 means off and 1 means on. $P_{de,i}$ is the output schedule of the ith diesel generator for the next 24 h. $P_{ba,j}$ is the charging/discharging schedule of the jth ETESS for the next 24 hours. $P_{pv,l}$ is the output schedule of the lth PV system for the next 24 h. $P_{wt,m}$ is the output schedule of the mth wind power system for the next 24 h. $u_{de,i}$ is the unit commitment of the ith diesel generator for the next 24 h. $u_{ba,j}$ is the unit commitment of the jth ETESS for the next 24 h. $u_{sdload,k}$ is the load switching schedule of the kth secondary load. $u_{ctload,p}$ is the load switching schedule of the pth interruptible load.

Since the time scale of PTESS's charge-discharge cycling is second and charge-discharge capacity is very small, there is no need to optimize the output scheduling of PTESS in day-ahead optimization.

3.2.2. Objective Function and Constraints

To maximize the utilization of RES under the condition of ensuring MG safety and stability and maintain system reliability, the objective function consist of operation cost F and decision penalty term C. The operation cost includes diesel generators operation cost, ETESS cost, RES system cost and load profit (electricity selling profit and interruptible load compensation). The decision penalty term includes penalty of abandoned solar and wind power and load outage. Since the objective function includes the penalty term above, the system will make the decisions of RES curtailment or load shedding only when the safety and stability constraints cannot be satisfied. Considering system power balance constraint, reserve capacity constraint and operation constraints of each DGs, a day-ahead optimization model of stand-alone MG is established as follows:

$$
\begin{aligned}
y(P,u) &= F_{load}(u_{load}) - F_{ba}(P_{ba}, u_{ba}) - F_{de}(P_{de}, u_{de}) - F_{re}(P_{pv}, P_{wt}) \\
&\quad + C_{dep}(P_{pv}, P_{wt}) + C_{lpsp}(u_{load})
\end{aligned}
\tag{2}
$$

$$
F_{de} = \sum_{t=1}^{T_1} \sum_{i=1}^{n_{de}} \left[(s_{de,start,i}(t) f_{de,start} + s_{de,down,i}(t) f_{de,down}) + f_{diesel}(P_{de,i}\Delta T) + g_{diesel}(P_{de,i}\Delta T) \right]
\tag{3}
$$

$$
F_{ba} = \sum_{t=1}^{T_1} \sum_{j=1}^{n_{ba}} f_{ba,oper} \left| P_{ba,j}(t) \right| \Delta T + f_{ba,inv} D_{ba,cyc}
\tag{4}
$$

$$
F_{re} = \sum_{t=1}^{T_1} \left(\sum_{l=1}^{n_{pv}} \left[f_{pv,oper} P_{pv,l}(t)\Delta T \right] + \sum_{m=1}^{n_{wt}} \left[f_{wt,oper} P_{wt,m}(t)\Delta T \right] \right)
\tag{5}
$$

$$
F_{load} = \sum_{t=1}^{T_1} \sum_{k=1}^{n_{load}} \left[f_{load,sale} P_{load,k}(t) - f_{load,cut} \Delta P_{cutload,k}(t) \right]
\tag{6}
$$

$$C_{dep} = \beta_{dep} \sum_{t=1}^{T_1} \left(\sum_{l=1}^{n_{pv}} [P_{pv,l}^{short-term}(t) - P_{pv,l}(t)] + \sum_{m=1}^{n_{wt}} [P_{wt,m}^{short-term}(t) - P_{wt,m}(t)] \right) \tag{7}$$

$$C_{lpsp} = \beta_{lpsp} \sum_{t=1}^{T_1} \left(\sum_{k=1}^{n_{sdload}} [(1 - u_{sdload,k}(t))P_{sdload,k}(t)] + \sum_{p=1}^{n_{ctload}} [(1 - u_{ctload,p}(t))P_{ctload,p}(t)] \right) \tag{8}$$

s.t.

$$\sum_{i=1}^{n_{de}} P_{de,i}(t) + \sum_{j=1}^{n_{ba}} P_{ba,j}(t) \qquad \sum_{s=1}^{n_{ipload}} P_{ipload,s}(t) + \sum_{p=1}^{n_{ctload}} u_{ctload,p}(t)P_{ctload,p}(t)$$
$$+ \sum_{l=1}^{n_{pv}} P_{pv,l}(t) + \sum_{m=1}^{n_{wt}} P_{wt,m}(t) \qquad = \qquad + \sum_{k=1}^{n_{sdload}} u_{sdload,k}(t)P_{sdload,k}(t) \tag{9}$$

$$u_{ba,j}(t)P_{ba,n,j}^{cha} \le P_{ba,j}(t) \le u_{ba,j}(t)P_{ba,n,j}^{dis} \tag{10}$$

$$Soc_{ba,min,j} \le Soc_{ba,j}(t) \le Soc_{ba,max,j} \tag{11}$$

$$Soc_{ba,j}(1) - \Delta Soc_{balance} \le Soc_{ba,j}(T_1) \le Soc_{ba,j}(1) + \Delta Soc_{balance} \tag{12}$$

$$u_{de,i}(t)\beta_{de,i,min}P_{de,n,i} \le P_{de,i}(t) \le u_{de,i}(t)P_{de,n,i} \tag{13}$$

$$-\Delta P_{de,down,i} \le P_{de,i}(t) - P_{de,i}(t-1) \le \Delta P_{de,up,i} \tag{14}$$

$$0 \le P_{pv,l}(t) \le P_{pv,l}^{short-term}(t) \tag{15}$$

$$0 \le P_{wt,m}(t) \le P_{wt,m}^{short-term}(t) \tag{16}$$

$$\sum_{i=1}^{n_{de}} \min(u_{de,i}(t)P_{de,n,i} - P_{de,i}(t), \Delta P_{de,up,i}) +$$
$$\sum_{j=1}^{n_{ba}} \min\left(u_{ba,j}(t)P_{ba,n,j}^{dis} - P_{ba,j}(t), \frac{E_{ba,n,j}(Soc_{ba,j}(t) - Soc_{ba,min,j})}{\Delta T}\eta_{ba,dis}\right) \ge R_s(t) \tag{17}$$

where F_{de} is the operation cost of diesel generators in a scheduling period T_1. T_1 is the time length of day-ahead scheduling, $T_1 = 24$ h. n_{de} is the number of diesel generators. ΔT is time interval, $\Delta T = 1$ h. $s_{de,start,i}(t)$ and $s_{de,down,i}(t)$ are the operating state switching variable of ith diesel generator in the tth time period, $s_{de,start,i}(t) = 1$ means the operating state of ith diesel generator switched from stopped to running. While $s_{de,down,i}(t) = 1$ means the operating state of ith diesel generator from running to stopped. $u_{de,i}(t)$ is the operating state of ith diesel generator in the tth time period, $u_{de,i}(t) = 1$ means the ith diesel generator in a running state in the tth time period, while $u_{de,i}(t) = 0$ means in a stopped state. $f_{de,start}$ and $f_{de,down}$ are the start-up cost and the shut-down cost of diesel generators. $f_{diesel}(\cdot)$ is the fuel cost function of diesel generators. $g_{diesel}(\cdot)$ is the converted environmental cost function of diesel generators. F_{ba} is operation cost of ETESS in a schedule length T_1. n_{ba} is the number of ETESS. $u_{ba,j}(t)$ is the operation state of the jth ETESS in the tth time period, $u_{ba,j}(t) = 1$ means that ETESS is in operation and $u_{ba,j}(t) = 0$ means that ETESS is in a shutdown state. $f_{ba,oper}$ is the operation cost coefficient of ETESS measured in yuan/kW. $P_{ba,j}(t)$ is the output power of the jth ETESS in the ith time period. $f_{ba,inv}$ is the initial investment cost of ETESS. $D_{ba,cyc}$ is the lifetime degradation ration in a schedule length T_1, $D_{ba,cyc}$ is estimated based on throughput counting model. $f_{wt,oper}$ and $f_{pv,oper}$ are the operation cost coefficients of WT system and PV system F_{load} is the load profit during the schedule length T_1. n_{load} is the number of load in MG. $f_{load,sale}$ is the electricity selling price. $f_{load,cut}$ is the coefficient of interruptible compensation. $P_{load,k}(t)$ is the power of the kth load in the tth time period. $\Delta P_{cutload,k}(t)$ is the reduction of the kth interruptible load in the tth time period. $P_{pv,l}^{short-term}$ and $P_{wt,l}^{short-term}$ are the short-term predicted power of PV system and WT system. β_{dep} and β_{lpsp} are the penalty coefficient of RES curtailment and load outage. n_{ipload} is the number of important load in the system. $P_{ipload,s}(t)$ is the power of the sth important load in the tth time period. $P_{ba,n,j}^{cha}$ and $P_{ba,n,j}^{dis}$ are the rated charging power and discharging power of the jth ETESS. $Soc_{ba,max,j}$ and $Soc_{ba,min,j}$ are the maximum and minimum limit of the jth ETESS SOC. $Soc_{ba,j}(1)$ and $Soc_{ba,j}(T_1)$ are the SOC value of the *j*th ETESS at the beginning and the ending of a schedule, $\Delta Soc_{balance}$ is the permitted deviation of ETESS energy balance in a cycle. $P_{de,n,i}$ is the rated power of *i*th diesel generator. $\beta_{de,i,min}$ is the minimum operating power factor of *i*th diesel generator. $\Delta P_{de,down,i}$ and $\Delta P_{de,up,i}$ are the maximum ramp down rate and ramp up rate of the *i*th diesel generator. $E_{ba,n,j}$ is the rated capacity of the *j*th ETESS. $\eta_{ba,dis}$ is the discharge efficiency of the ETESS. $R_s(t)$ is the reserve capacity requirement of MG.

Among the above constraints, Equation (9) is the system power balance equality constraint, Equation (10) is the ETESS operating power inequality constraint, Equation (11) is the ETESS SOC inequality constraint, Equation (12) is the ETESS daily cycle balance inequality constraint, Equation (13) is the DSG operating power constraint, Equation (14) is the DSG ramp rate inequality constraint, Equation (15) is the PV system operating power inequality constraint, Equation (16) is the wind system operating power inequality constraint, Equation (17) is the system reserve capacity constraint.

From the mathematical description formulas Equations (2)–(17), the day-ahead optimization model is a linear programming problem, so it is convex and has a unique global optimal solution.

3.3. Intraday Rolling Optimization Based on Model Predictive Control

The day-ahead optimization decisions are mainly based on the day-ahead short-term forecast results of wind power and PV system, so the accuracy of the model is heavily dependent on the accuracy of the forecast. Since the accuracy of intraday ultra-short-term forecast is higher than that of day-ahead short-term forecast, rolling correction of the day-ahead schedule is required.

3.3.1. Model Predictive Control Framework

Rolling scheduling in fact is an online rolling optimization problem over a short period time, the whole scheduling time is divided into several scheduling periods and an optimal scheduling strategy is solved online in each scheduling period Three characteristics of predictive control makes it suitable for rolling scheduling:

(1) Prediction model: Based on the measured output data of generators and the predictive data of RES and load data in the next time period, this model is used to predict the future output of generators.
(2) Rolling optimization: Repeated online optimization negates the influence caused by uncertain factors such as RE fluctuations.
(3) Feedback control: The measured data feedback is used to correct the actual output state of the generators in the model and ensure that the next optimization calculation is based on the latest measured data.

Figure 3 is the framework of MPC (model predictive control, MPC) based rolling scheduling optimization framework. Use the measured state feedback and prediction data to solve an optimal control sequence. For example, the following equation is the optimal control sequence:

$$U^* = [u^*(t|t), u^*(t+1|t), \cdots, u^*(t+k-1|t)] \tag{18}$$

where *k* is the prediction horizon. U^* is the optimal control sequence for the following *k* time periods solved at time. $u^*(t|t)$ is the optimal control rule for the first time period solved at time *t*, comprised of a vector of control variables. In this paper, the rolling time window length, that is, the prediction horizon, is $k = 4$ h, time interval is $\Delta t = 15$ min.

3.3.2. Decision Variables

Decision variables in this model is almost the same as that of the day-ahead optimization model with a slight difference of variable dimensions.

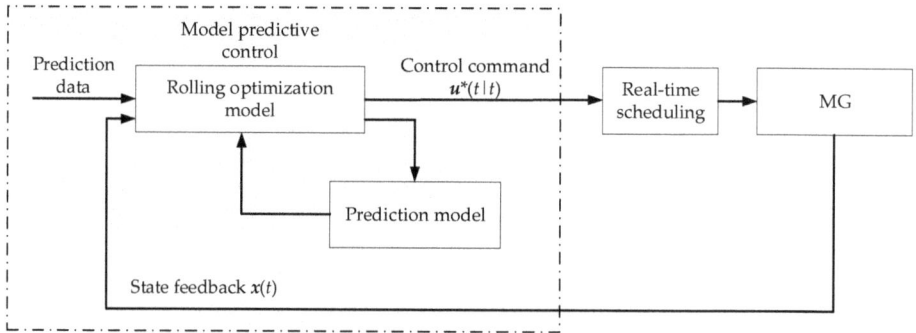

Figure 3. MPC based rolling scheduling optimization framework.

3.3.3. Objective Function and Constraints

In the traditional intraday rolling scheduling optimization mode for grid-connected MG, the mode only corrects the scheduling for DGs but not the unit commitment and load switching. Besides, for MG with high RE penetration, the randomness and the intermittent of sources and load have negative effect on stable operation of MG. Thus, in this intraday rolling scheduling optimization model, optimization for unit commitment and load switching is required. Also, the SOC of ETESS is expected to follow the day-ahead scheduling as possible. So, the objective function also consists of operation cost term F and decision regularization term C. F is basically the same as that in the day-ahead optimization model while C includes SOC deviation correction penalty of ETESS, PV and WT curtailment penalty, load switching scheduling correction penalty and unit commitment correction penalty of diesel generators. The objective function is described as follows:

$$
\begin{aligned}
y(P, u) &= F_{\text{load}}(u_{\text{load}}) - F_{\text{ba}}(P_{\text{ba}}, u_{\text{ba}}) - F_{\text{de}}(P_{\text{de}}, u_{\text{de}}) - F_{\text{re}}(P_{\text{pv}}, P_{\text{wt}}) \\
&+ C_{\text{dep}}(P_{\text{pv}}, P_{\text{wt}}) + C_{\text{crt\_soc}}(P_{\text{ba}}, u_{\text{ba}}) + C_{\text{crt\_ld}}(u_{\text{load}}) + C_{\text{crt\_de}}(u_{\text{de}})
\end{aligned}
\tag{19}
$$

where $C_{\text{crt\_soc}}(P_{\text{ba}}, u_{\text{ba}})$, $C_{\text{crt\_ld}}(u_{\text{load}})$, $C_{\text{crt\_de}}(u_{\text{de}})$ are the SOC error correction regularization term, load switching schedule correction regularization term and unit commitment correction regularization term. The meanings of other variables are already described in the day-ahead optimization objective function. This model executes every 15 min, obtaining the schedule results for the following 4 h and take the first 15 min result as the MPC decision command. The newly added regularization terms equations are described as follows:

$$
C_{\text{crt\_soc}} = \beta_{\text{crt\_soc}} \sum_{t=1}^{T_2} \left(\sum_{j=1}^{n_{\text{ba}}} [Soc_{\text{ba},j}(t) - Soc_{\text{ba},j}^{\text{Day\_ahead}}(t)] \right)
\tag{20}
$$

$$
C_{\text{crt\_ld}} = \beta_{\text{crt\_ld}} \sum_{t=1}^{T_2} \left(\sum_{k=1}^{n_{\text{sdload}}} \text{abs}(u_{\text{sdload},k}(t) - u_{\text{sdload},k}^{\text{Day\_ahead}}(t)) + \sum_{p=1}^{n_{\text{ctload}}} \text{abs}(u_{\text{ctload},p}(t) - u_{\text{ctload},p}^{\text{Day\_ahead}}(t)) \right)
\tag{21}
$$

$$
C_{\text{crt\_de}} = \beta_{\text{crt\_de}} \sum_{t=1}^{T_2} \left(\sum_{i=1}^{n_{\text{de}}} \text{abs}(u_{\text{de},i}(t) - u_{\text{de},i}^{\text{Day\_ahead}}(t)) \right)
\tag{22}
$$

where $\beta_{\text{crt\_soc}}$, $\beta_{\text{crt\_ld}}$, $\beta_{\text{crt\_de}}$ are the coefficients of SOC error correction regularization, load switching schedule correction regularization and unit commitment correction regularization. T_2 is the rolling window length, which is usually set to 4 h.

3.4. Comprehensive Criteria Based Quasi-Real-Time Coordinated Control

Although the prediction accuracy of RES and load have improved greatly in intraday rolling optimization model, errors still exist. The intraday rolling optimization model in this paper is performed once every 15 min and at the interval between two execution times, the unbalanced power caused by the prediction error should not be completely borne by the diesel generator (when it is the main source). Meanwhile, in order to improve the ability of ETESS actual SOC to follow the rolling optimization results, quasi-real-time unbalanced power can be used to correct the charge/discharge power of ETESS under some conditions. Therefore, in order to satisfy the mentioned requirements and further deal with the prediction error, this paper adds a control process with smaller timescale.

To improve calculation speed and reduce decision making time, a quasi-real-time coordinated control criterion system is developed in this paper, including quasi-real-time unbalanced power, power command of ETESS in rolling optimization and quasi-real-time SOC of ETESS. Different quasi-real-time coordinated control strategies based on different criteria states, as shown in Table 1.

Table 1. Quasi-real-time coordinated control criterion system.

Quasi-Real-Time Power Imbalance	Power of ETESS in Rolling Optimization	Quasi-Real-Time SOC of ETESS
① $P_{ubl}^m(t) > 0$	③ $P_{ba}^h(t) > 0$	⑤ $Soc_{ba}^m(t) > Soc_{ba}^h(t)$
② $P_{ubl}^m(t) < 0$	④ $P_{ba}^h(t) \leq 0$	⑥ $Soc_{ba}^m(t) \leq Soc_{ba}^h(t)$

where $P_{ubl}^m(t)$ is the measured quasi-real-time power imbalance value. $P_{ba}^h(t)$ is the command value of rolling optimization for ETESS power. $Soc_{ba}^m(t)$ is the measured quasi-real-time SOC value of ETESS. $Soc_{ba}^h(t)$ is the scheduled SOC value of ETESS given by rolling optimization.

The quasi-real-time unbalanced power determines the type of DGs that can be used to adjust the unbalanced power and the adjustment direction of the power increment. When the quasi-real-time unbalanced power is greater than zero, it indicates that there is a power shortage in MG system and diesel generators and ETESS is determined to participate in the adjustment of unbalanced power. Conversely, when the quasi-real-time unbalanced power is less than zero, it indicates that there is a power surplus in MG system, all types of DGs can take part the adjustment of unbalanced power.

The ETESS power command value determined from rolling optimization can provide criteria for the decision of ETESS quasi-real-time charge/discharge state which is expected to follow the command from rolling optimization. Besides, the quasi-real-time SOC measurement of ETESS can also provide criteria for the decision of ETESS quasi-real-time charge/discharge state. When the quasi-real-time SOC measurement of ETESS is greater than the SOC determined from rolling optimization, the decision of ETESS is expected to increase the discharge power or reduce the charge power, so that SOC of ETESS is close to the scheduled value as soon as possible. Likewise, when the measurement is less than the scheduled value, it is determined to reduce discharge power or increase charge power.

The 8 types of control strategies can be obtained by arranging and combining the above criteria, and the adjustment priorities of DGs in different types of control strategies may be different. When the quasi-real-time unbalanced power is less than zero, if the DER with highest priority cannot fully compensate the unbalanced power, PV or WT system will reduce output power until it is fully compensated.

(1) When ①③⑤ are satisfied, increase the discharge power of ETESS with high priority. The increment $\Delta P_{ba}^m(t)$ is calculated by SOC error and rated capacity, as described by the following equation:

$$\Delta P_{ba}^m(t) = \min\left(P_{ba,n}^{dis} - P_{ba}^h(t), \frac{E_{ba,n}(Soc_{ba}^m(t) - Soc_{ba}^h(t))\eta_{ba,dis}}{\Delta T_{rest}}, P_{ubl}^m(t)\right) \tag{23}$$

where $\Delta T_{rest}(t)$ is the resting time until next rolling optimization.

(2) When ①③⑥ are satisfied, increase the output of diesel generators with high priority. The increment $\Delta P_{de}^m(t)$ takes rated power and spinning reserve margin into consideration, as described by the following equation:

$$\Delta P_{de}^m(t) = \min(P_{de,n} - R_{de}(t) - P_{de}^h(t), P_{ubl}^m(t)) \tag{24}$$

where $P_{de}^h(t)$ is the command of rolling optimization for diesel generators and $R_{de}(t)$ is the spinning reserve margin.

(3) When ①④⑤ are satisfied, decrease the charge power of ETESS with high priority. $\Delta P_{ba}^m(t)$ is described by the following equation:

$$\Delta P_{ba}^m(t) = \min(-P_{ba}^h(t), P_{ubl}^m(t)) \tag{25}$$

(4) When ①④⑥ are satisfied, the strategy is the same as Equation (2).
(5) When ②③⑤ are satisfied, decrease the output of diesel generators with high priority. The increment $\Delta P_{de}^m(t)$ needs to take minimum load limitation into consideration, as described by the following equation:

$$\Delta P_{de}^m(t) = \max(\beta_{de,min} P_{de,n} - P_{de}^h(t), P_{ubl}^m(t)) \tag{26}$$

(6) When ②③⑥ are satisfied, decrease the discharge power of ETESS with high priority. $\Delta P_{ba}^m(t)$ is described by the following equation:

$$\Delta P_{ba}^m(t) = -\min(P_{ba}^h(t), P_{ubl}^m(t)) \tag{27}$$

(7) When ②④⑤ are satisfied, the strategy is the same as Equation (5).
(8) When ②④⑥ are satisfied, increase the charge power of ETESS with high priority. $\Delta P_{ba}^m(t)$ is described by the following equation:

$$\Delta P_{ba}^m(t) = \max(P_{ba,n}^{cha} - P_{ba}^h(t), \frac{E_{ba,n}(Soc_{ba}^m(t) - Soc_{ba}^h(t))\eta_{ba,cha}}{\Delta T_{rest}}, P_{ubl}^m(t)) \tag{28}$$

3.5. Real-Time Correction Control of Hybrid Energy Storage

Since the time scale of real-time correction control is short, DSGs with low response speed such as PV and WT are not incorporated into this process. Therefore, real-time correction process of MGCC only considers HESS remote adjustment. During the process of real-time coordinated control model, HESS is used to compensate second-level power imbalance. ETESS compensates the slowly varying component in power imbalance, while PTESS compensates the fast-varying component. Besides, to solve the SOC limit violation problem, a PTESS energy calibration method is developed to ensure the PTESS can maintain the ability to charge/discharge continuously.

This paper uses the first order low-pass filter (FOLPF) algorithm to calculate the component power for PTESS. The transfer function of FOLPF algorithm can be described as follows:

$$P_{nbl}^{sf}(s) = \frac{1}{1 + sT_{stab}} P_{nbl}^s(s) \tag{29}$$

where $P_{nbl}^s(s)$ $P_{nbl}^{sf}(s)$, are the power imbalance and the Laplace transform of its low-frequency component. T_{stab} is the time constant of low-pass filter; s is the Laplacian operator.

If s is substituted by d/dt and use Δt as the calculation step, then the equation of FOLPF in time domain can be described as follows:

$$P_{\text{nbl}}^{\text{sf}}(t) = \frac{T_{\text{stab}}}{\Delta t + T_{\text{stab}}} P_{\text{nbl}}^{\text{sf}}(t - \Delta t) + \frac{\Delta t}{\Delta t + T_{\text{stab}}} P_{\text{nbl}}^{\text{s}}(t) \tag{30}$$

where $P_{\text{nbl}}^{\text{sf}}(t)$, $P_{\text{nbl}}^{\text{sf}}(t - \Delta t)$ are the low-frequency component of power imbalance at the current time period and the last time period; $P_{\text{nbl}}^{\text{s}}(t)$ is the power imbalance at the current time period; Δt is the data sampling interval, which is the timescale of real-time coordinated control, $\Delta t = 5$ s.

ETESS is responsible for dealing with the lower frequency component $\Delta P_{\text{ba}}^{\text{s}}(t)$, while the high-frequency component is dealt with by PTESS. The real-time coordinated control command for ETESS and PTESS are:

$$P_{\text{ba}}^{\text{s}}(t) = P_{\text{ba}}^{\text{s}}(t - \Delta t) + \Delta P_{\text{ba}}^{\text{s}}(t) = P_{\text{ba}}^{\text{s}}(t - \Delta t) + P_{\text{nbl}}^{\text{sf}}(t) \tag{31}$$

$$P_{\text{sc}}^{\text{s}}(t) = P_{\text{nbl}}^{\text{s}}(t) - P_{\text{ba}}^{\text{s}}(t) \tag{32}$$

Considering the ETESS is expected to follow intraday rolling optimization schedule, the time constant of FOLPF should be able to adjust adaptively so that the ETESS can compensate the low-frequency component of power imbalance and make its SOC close to the rolling optimization result at the same time. To achieve this, a rule for adjusting the time constant in FOLPF is developed:

$$T_{\text{stab}}(t) = [1 - \lambda(t) \cdot (Soc_{\text{ba}}^{\text{s}}(t) - Soc_{\text{ba}}^{\text{h}}(t))] T_{\text{stab}}^{\text{ref}} \tag{33}$$

$$\lambda(t) = \text{sign}(P_{\text{nbl}}^{\text{s}}(t)) \cdot \lambda_{\text{T}}^{\text{ref}} \tag{34}$$

where $T_{\text{stab}}^{\text{ref}}$ is the reference for time constant; $Soc_{\text{ba}}^{\text{s}}(t)$ is the SOC of ETESS measured in real-time; $\lambda(t)$ is the adjusting coefficient; $\lambda_{\text{T}}^{\text{ref}}$ is the reference for the adjusting coefficient; $\text{sign}(\cdot)$ is the sign function, whose value is 1 if the variable is positive, -1 if negative and 0 if the variable is 0.

When power imbalance $P_{\text{nbl}}^{\text{s}}(t)$ is greater than 0, $\lambda(t)$ is a positive value. If ETESS SOC is also greater than the rolling optimization value, then $T_{\text{stab}}(t) < T_{\text{stab}}(t)$ and the power adjustment $\Delta P_{\text{ba}}^{\text{s}}(t)$ dispatched to ETESS will increase, causing the SOC to decrease rapidly to the optimal value. But if the SOC is less than the rolling optimization value, then $\Delta P_{\text{ba}}^{\text{s}}(t)$ will decrease, slowing down the offset velocity of SOC. Same effects are expected when $P_{\text{nbl}}^{\text{s}}(t)$ is less than 0.

But when the SOC of PTESS is too high or too low, positive and negative high-frequency power imbalance cannot be compensated at the same time, requiring blocking PTESS's charging function or discharging function. If the SOC of PTESS is too low, the discharging function is blocked and PTESS only compensates the negative high-frequency power imbalance, while the positive high-frequency component is compensated by diesel generators. If the SOC of PTESS is too high, the charging function is blocked and PTESS only compensates the positive high-frequency power imbalance, while the negative high-frequency component is compensated by diesel generators. When the PTESS's SOC is back around the middle value, the charge/discharge function is unblocked.

4. Case Study

This paper uses the topology shown in Figure 1 as the study case. This MG contains two diesel generators with rated power 50 kW, a wind power system with rated power 60 kW, a PV system with rated power 70 kW, a HESS consists of lead acids batteries (50 kW/200 kWh) and SC (50 kW/ 5 kWh) and three loads (load 1 is an important load, 2 and 3 are secondary loads). Simulation software Matlab R2014a is used and ILOG CPLEX solver is used in day-ahead and intraday model. Our code is available upon request.

4.1. Basic Data

The basic data analyzed in this example includes the parameters of each power generation unit in the MG, multi-time scale coordinated control related parameters, the short-term power prediction results, ultra-short-term power prediction and real-time power data of WT systems, PV systems and loads. The parameters of DSGs, RES, HESS, and multi-timescale coordinated control models are as shown in Tables 2–6.

Table 2. Parameters of diesel generator.

Serial Number of DSG	Rated Power (kW)	Startup Cost (CNY/Per Time)	Shutdown Cost (CNY/Per Time)	Coefficient of Operation Cost (CNY/kWh)	Fuel Cost (CNY/kWh)	Fuel Curve Slope (L/kWh)	Minimum Load Factor (%)
1	50	2	2	0.0088	7	0.	30
2	50	2	2	0.0088	7	0.3	30

Table 3. Parameters of renewable energy sources.

Type	Rated Power (kW)	Coefficient of Operation Cost (CNY/kWh)
Wind power	50	0.0296
PV power	50	0.0096

Table 4. Parameters of HESS.

Type	Rated Power/Capacity (kW/kWh)	Initial Investment Cost (CNY/kWh)	Coefficient of Operation Cost (CNY/kWh)	Energy Per Unit Capacity (kWh)	Permitted SOC Range (%)	SOC Warning Range (%)	Initial SOC (%)
ETESS	50/200	2	0.0088	250	[10, 90]	[30, 70]	50
PTESS	50/5	11.4	0	∞	[10, 90]	[30, 70]	50

Table 5. Parameters of multi-timescale coordinated control (day-ahead and intraday model).

Energy Balance Error in a Cycle (%)	DSG Minimum Operation (Shutdown) Time (h)	Reserve Capacity (kW)	Regularization Parameter of RE Curtailment	Regularization Parameter of Power Shortage	Regularization Parameter of DSG Unit Commitment Correction	Regularization Parameter of SOC Error Correction
10	2	10	5000	10000	10000	500

Table 6. Parameters of multi-timescale coordinated control (Quasi-real-time and real-time model).

Adjusting Coefficient for Time Constant of FOLPF	Reference Value for Time Constant of FOLPF
15	30

Short-term and ultra-short-term power prediction results are shown in Figure 4:

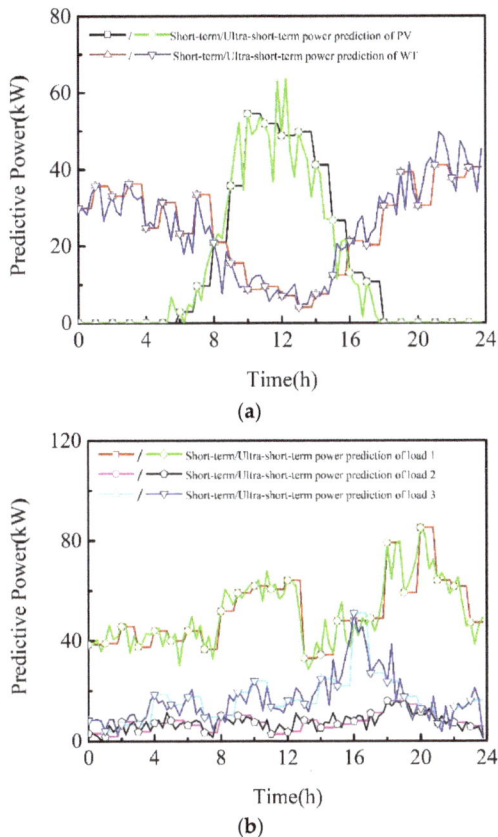

Figure 4. Short-term and ultra-short-term power prediction results. (**a**) Prediction results of RES. (**b**) Prediction results of load.

4.2. Analysis of Multi-Timescale Coordinated Control

4.2.1. Analysis of Day-Ahead Optimization

The day-ahead scheduling results are shown in Figure 5. The startup and shutdown schedule of two DSGs are: DSG1 runs during 0:00–23:00 and DSG2 runs during 16:00–21:00. Because during 16:00–21:00, the predicted load is increasing and the predicted RES power is decreasing (due to the fading of sunlight), DSG2 is activated to generate power for the load within the MG. During 0:00–23:00, at least one DSG is running as the main power supply for the net load. When the net load is less than the minimum load factor of DSG, ETESS starts to charge to ensure DSG operates normally. After 21:00, the predicted net load starts to decrease and ETESS needs to discharge to make its SOC value back to around 0.5, so DSG2 needs to be shut down during 21:00–24:00 and as for DSG1, 23:00–24:00. Therefore, ETESS is the main power supply for the MG after 23:00 and its SOC value drop backs to around 0.5.

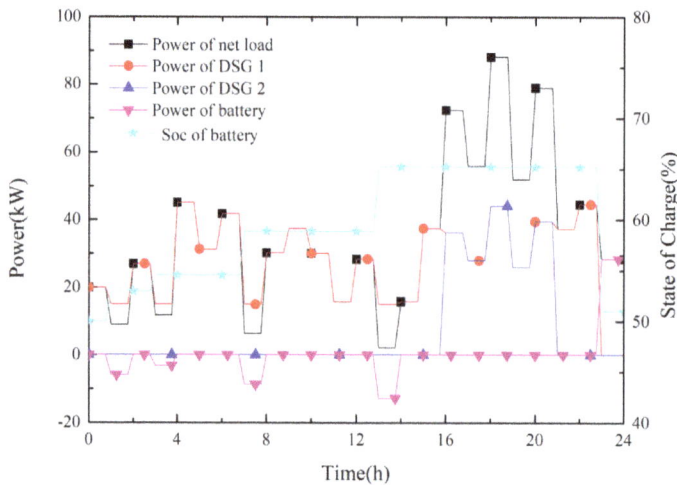

Figure 5. Day-ahead schedule of each type of DG.

In the day-ahead optimization results, there is no need to shed load and limit the power of RES, so the corresponding regularization term is 0. To verify the effects of RES curtailment and power shortage regularization term, a comparison test is setup. In the comparison test, RES curtailment and power shortage regularization term are neglected and the other conditions are kept the same. The day-ahead optimization results of the comparison test are shown in Table 7. As the results suggest, without regularization terms in the day-ahead optimization model, the power shortage percentage can reach as high as 28.3567% and the system operation cost is 62.57% less than the cost when regularization is taken into account. These results show that generation cost of standalone MG is still high. Even when the electricity selling price is 1.5 CNY, the marginal generation cost is still higher than marginal usage cost. Therefore, for stand-alone MG that needs to ensure the reliability of power supply, it is necessary to sacrifice some economic benefits to reduce the load power shortage. An effective way is to add power shortage regularization term in day-ahead optimization model.

Table 7. Comparison test for the verification of effects of regularization terms in day-ahead optimization model.

Day-Ahead Optimization Results	Without Regularization Terms	With Regularization Term
RE curtailment (%)	0.00095619	0
Power shortage (%)	28.3567	0
MG operation cost (CNY)	−1049.9027	−645.8309

4.2.2. Analysis of Intraday, Quasi-Real-Time and Real-Time Coordinated Control

Since intraday rolling optimization, quasi-real-time and real-time coordinated control all need to use measured data as the input of the control algorithm, the coordinated control strategy decisions of these three timescales are strongly coupled. that is, the control result of shorter timescale will affect the decision of longer timescale. Therefore, the case analysis of these three parts are discussed together in this section. Figure 6 shows the results (0–24 h) of multi-timescale coordinated control case study.

As Figure 6 shows, the short-term and ultra-short-term forecasts have small errors, so there is no need to correct the startup and shutdown plan of DSGs in intraday rolling optimization. Under the condition that the plan follows the day-ahead scheduling result, observe the changing pattern of the operating power of DSG in various timescales:

(1) The essential of seconds-scale real-time coordinated control is to use HESS to compensate the power imbalance in the interval of quasi-real-time coordinated control, so the power curve of DSG in real-time coordinated control is basically overlapped with that in quasi-real-time coordinated control.

(2) The power curve of DSG is more accurate in the shorter timescale. Compared to real-time power curve, the mean error of quasi-real-time curve is 0.19% while the mean error of intraday rolling curve is 11.73% and for day-ahead scheduling, 26.09%.

Since the charging and discharging time of the PTESS is short, usually in the scale of seconds, the HESS only controls the ETESS in the day-ahead, intraday, quasi-real-time timescale, while the PTESS is only coordinated in real-time timescale to suppress the high frequency unbalanced power. Observe the changing characteristics of the operating power of ETESS in various timescales:

(1) ETESS needs to compensate the low frequency unbalanced power in real-time timescale, so its real-time operating power curve is slightly different from the quasi-real-time curve.

(2) Though intraday and day-ahead curves differ from the real-time curve greatly, they still reflect the overall change of the energy of ETESS.

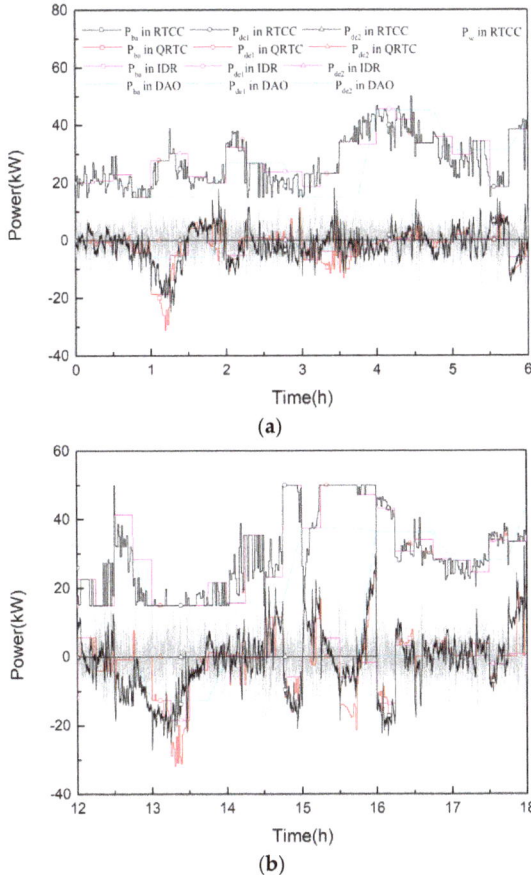

Figure 6. *Cont.*

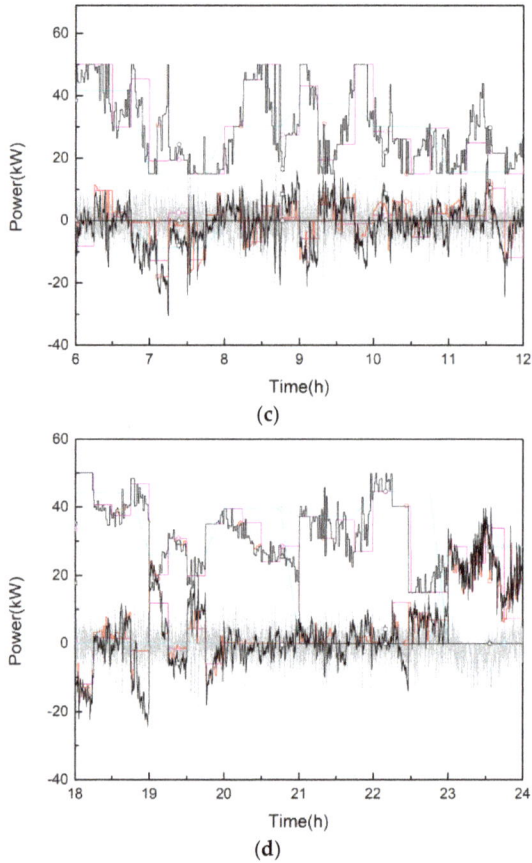

Figure 6. Test results of multi-timescale coordinated control. (a) Results of 0–6h. (b) Results of 6–12h. (c) Results of 12–18h. (d) Results of 18–24h.

Figure 7 shows the SOC curve of ETESS in various timescales and suggests that the real-time SOC curve can follow the day-ahead schedule accurately. This is due to introducing SOC error correction regularization term in intraday rolling model, comprehensive criteria in quasi-real-time model and adaptively adjusted time constant of FOLPF in real-time model. It can be seen from this figure that PTESS charge and discharge frequently to compensate the high frequency unbalanced power. If ETESS is used instead, the frequent charge and discharge activity will greatly reduce its cycle lifetime. During 23:00−24:00, DSGs are shut down and ETESS becomes the main power supply and its SOC is higher than the day-ahead schedule, so the system needs to compensate positive unbalanced power. The time constant of FOLPF decreases and the duty to compensate positive high frequency power imbalance shifts from PTESS to ETESS, making the SOC of ETESS falls to the scheduled value and the SOC if PTESS rises up quickly.

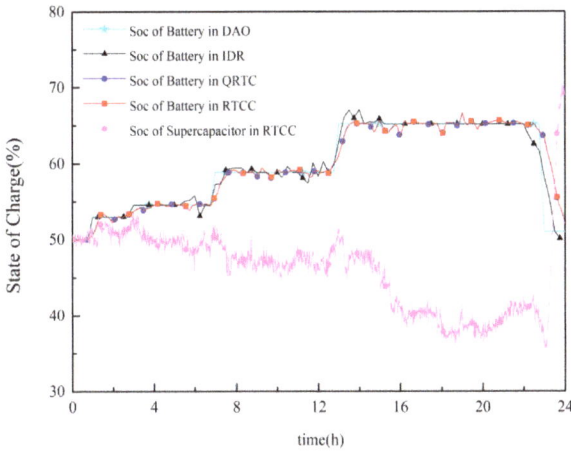

Figure 7. SOC curve of ETESS in various timescales.

4.2.3. Analysis of Daily SOC Balance of ETESS

The SOC of ETESS being balanced at the beginning and end of the daily cycle is one the factors that MG can operate stably. To achieve this goal, this paper makes some improvements in four aspects: introduce daily SOC balance constraint of ETESS in day-ahead scheduling. Add an SOC error regularization term in the objective function of intraday rolling optimization. Use comprehensive criteria to decide the priority of dispatching each DG in quasi-real-time coordinated control, further reducing the SOC error. Use adaptively adjusted time constant in FOLPF to adjust the SOC of ETESS. To analyze the effects of the above methods on daily SOC balance, a comparison test is conducted and the results are shown in Figure 8:

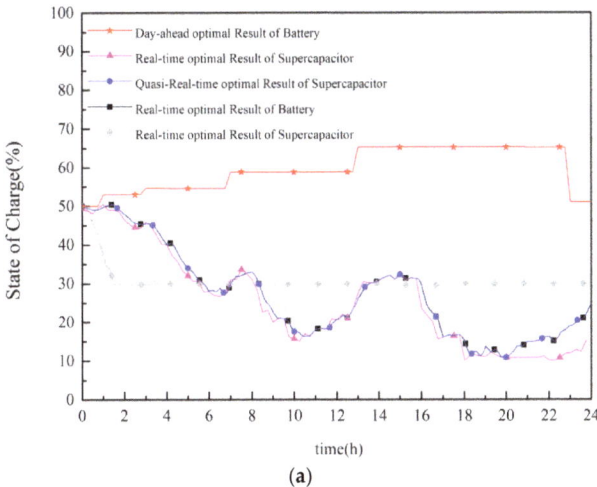

(a)

Figure 8. *Cont.*

(b)

(c)

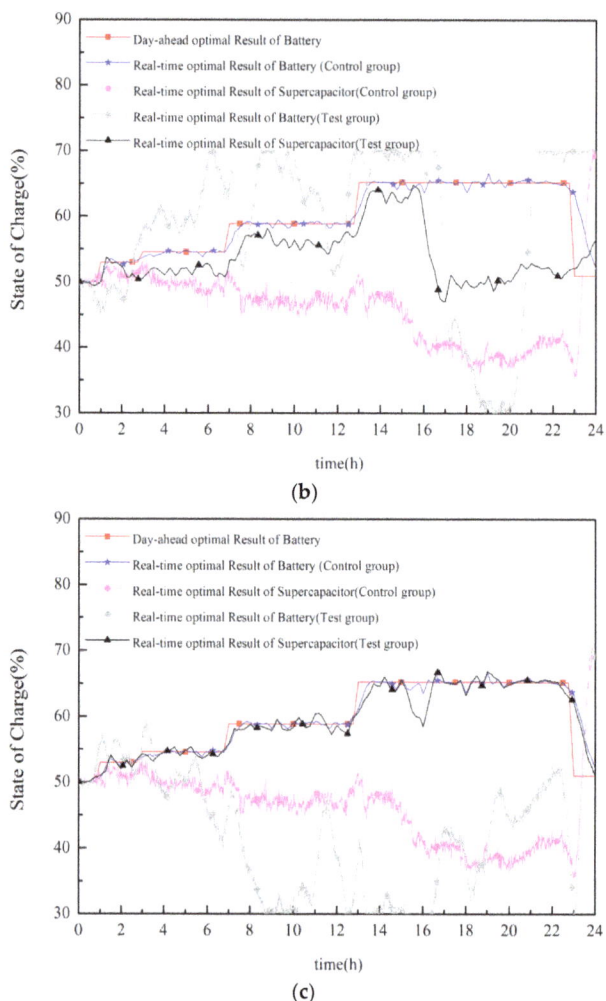

Figure 8. SOC waveforms of the HESS. (**a**) Ignore the SOC error regularization in intraday rolling. (**b**) Adjust ETESS with high priority in quasi-real-time coordinated control and does not use comprehensive criteria. (**c**) Adjust DSGs with high priority in quasi-real-time coordinated control and does not use comprehensive criteria.

Figure 8a shows the SOC waveforms when ETESS SOC error regularization is ignored in intraday rolling optimization. In this case, the ETESS SOC does not follow the day-ahead schedule and the value even drops below 30% in period 24:00. Meanwhile, because the ETESS SOC stays lower than the day-ahead schedule, when the seconds-timescale power imbalance is negative, time constant in FOLPF decrease adaptively, dispatching more negative unbalanced power to ETESS and less to PTESS, making PTESS SOC stays at a very low level.

Figure 8b shows the SOC waveforms when directly prioritize the adjustment of ETESS instead of considering comprehensive criteria in quasi-real-time coordinated control. The figure suggests the ETESS SOC error at the beginning and end of the day period is small but the SOC curve fails to follow the day-ahead schedule. This is because both positive and negative unbalanced power are compensated by ETESS. Since the ETESS SOC will deviate from the day-ahead schedule when power imbalance is very great. Besides, adjustment of time constant of FOLPF is only applicable in SOC fine-tuning but not large deviation.

Figure 8c shows the SOC waveforms when directly prioritize the adjustment of DSGs instead of considering comprehensive criteria in quasi-real-time coordinated control. The real-time ETESS SOC can follow the day-ahead schedule but due to the lack of adjustment of ETESS SOC in quasi-real-time coordinated control, the ETESS SOC regulation pressure is larger in real-time coordinated control, increasing the risk of PTESS's energy exceeding the limit.

4.2.4. Analysis of the DSG Unit Commitment Correction Regularization

To verify the effects of the DSG unit commitment correcting regularization, this study modifies the day-ahead load prediction data to increase its prediction error, making the prediction data during 15:30–16:00 far less than the actual predicted data and the real-time data. The day-ahead and intraday unit commitment are shown in Figure 9.

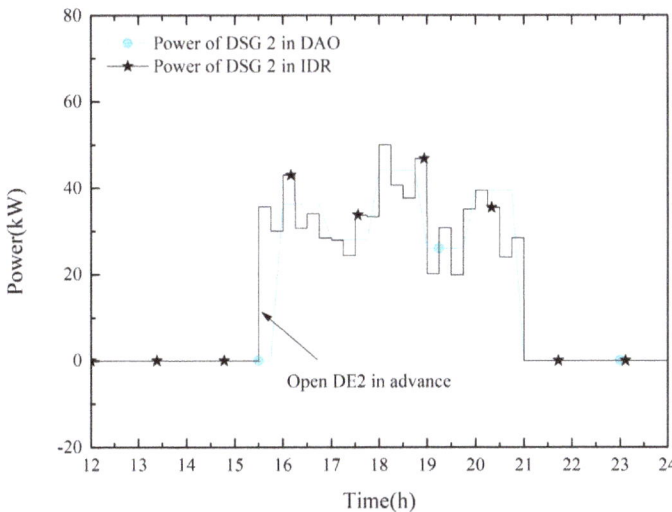

Figure 9. The DSG unit commitment.

Figure 9 shows when the prediction errors are so large that the system stability is threatened, the soft constraint which requires the unit commitment to follow the day-ahead schedule in intraday rolling optimization model, will lose its effect. The problem can be addressed by bringing the day-ahead schedule 0.5 h forward.

4.2.5. Analysis of the Necessity of Introducing Quasi-Real-Time Coordinated Control

The quasi-real-time coordinated control strategy is based on fixed logic rules and executes every minute. Introducing this model is to provide a transitional 1-min-timescale control between 15-min-timescale and 5-s-timescale control, smoothing the fluctuation by adjusting the power command of source and energy storage. Besides, the quasi-real-time coordinated control model uses comprehensive criteria to prioritize the dispatch of DGs, ensuring the ETESS SOC can follow the day-ahead schedule as possible. To verify this effect, a comparison test is set up, ignoring the quasi-real-time model while other conditions are kept the same. The test results are shown in Figure 10.

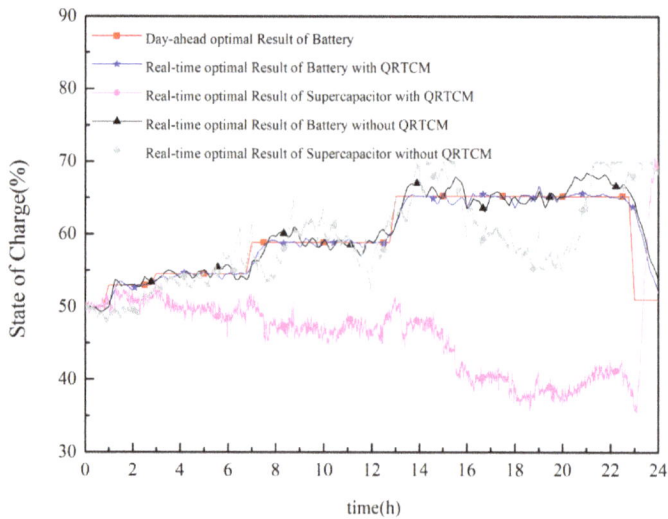

Figure 10. SOC comparison between situations whether using quasi-real-time model.

It can be seen from Figure 10 that ETESS SOC can follow the day-ahead schedule but has larger fluctuation when quasi-real-time model is ignored. In a MG that has small ETESS capacity, the risk of energy exceeding limit is greater. Besides, the PTESS SOC exceeds the limit multiple times because without the 1-min-timescale power imbalance compensation process, HESS needs to compensate unbalanced power every 5 s in a 15 min period. With such great regulating pressure, comes with the great risk of PTESS SOC exceeding the limit. In other words, introducing quasi-real-time coordinated control can enable ETESS SOC to follow the day-ahead schedule more accurately and reduce the operation risk of HESS.

5. Conclusions

This paper proposes a multi-timescale coordinated control scheme for stand-alone MG, consisting of "day-ahead optimization + intraday rolling + quasi-real-time correction + real-time coordinated control." The day-ahead model uses economic benefit as objective to schedule unit commitment and load switching. In the intraday rolling model, three regularization terms are introduced as soft constraint, which are the SOC error correction regularization term, load switching schedule correction regularization term and unit commitment correction regularization term. In the quasi-real-time model, comprehensive criteria are used to prioritize the adjustment of every DG. In the real-time model, an improved FOLPF algorithm is used to decide the power command for HESS.

From the case study analysis, the following conclusions can be drawn:

(1)   Introduce power shortage regularization term in day-ahead optimization model can significantly improve the power supply reliability but at the cost of some economic profit.

(2)   Based on ultra-short-term forecasting, which has smaller prediction error, the intraday rolling optimization model can adjust the day-ahead schedule for DGs output power appropriately. When the day-ahead prediction error is large, DG unit commitment and load switching schedule will be correct to the robustness of the MG system.

(3)   To keep the ETESS SOC balance at the beginning and end of the day period, it is necessary to add ETESS SOC error regularization term in the intraday rolling optimization model. Besides, the comprehensive criterial proposed in quasi-real-time model can improve the ETESS's ability to follow the day-ahead schedule and reducing the HESS operations risk at the same time.

In the future, further research can be conducted in these aspects:

(1) To overcome the shortcomings of large error in day-ahead power prediction and increase the credibility of the optimization results, uncertainty optimization theory can be introduced, such as robust optimization, fuzzy programming and stochastic programming.

(2) In large scale stand-alone MG system, it may contain multiple HESSs. How to manage the charge and discharge commands and the SOC of each HESS in quasi-real-time and real-time timescale, is a problem to be solved.

Author Contributions: J.C., P.Y. and J.P. contributed to the conception of the study and the algorithm. Y.H. and Y.C. analyzed the data. J.C. and J.P. wrote this paper collectively.

Funding: This research was funded by the Technologies Planning Program of Guangdong Province, and grant number are 2016B020245001.

Acknowledgments: This work was supported by the Technologies Planning Program of Guangdong Province (2016B020245001).

Conflicts of Interest: The authors declare no conflict of interest.

References

1. Hartono, B.S.; Budiyanto; Setiabudy, R. Review of microgrid technology. In Proceedings of the 2013 International Conference on QiR, Yogyakarta, Indonesia, 25–28 June 2013.
2. Lasseter, R.H.; Paigi, P. Microgrid: A conceptual solution. In Proceedings of the 2004 IEEE 35th Annual Power Electronics Specialists Conference, Aachen, Germany, 20–25 June 2004.
3. He, M.; Giesselmann, M. Reliability-constrained self-organization and energy management towards a resilient microgrid cluster. In Proceedings of the 2015 IEEE Power & Energy Society Innovative Smart Grid Technologies Conference (ISGT), Washington, DC, USA, 18–20 February 2015.
4. Hatziargyriou, N.; Asano, H.; Iravani, R.; Marnay, C. Microgrids. *IEEE Power Energy Mag.* **2007**, *5*, 78–94. [CrossRef]
5. Dougal, R.A.; Liu, S.; White, R.E. Power and life extension of battery-ultracapacitor hybrids. *IEEE Trans. Compon. Packag. Technol.* **2002**, *25*, 120–131. [CrossRef]
6. Cagnano, A.; Bugliari, A.C.; De Tuglie, E. A cooperative control for the reserve management of isolated microgrids. *Appl. Energy* **2018**, *218*, 256–265. [CrossRef]
7. Lasseter, R.H. Microgrids. In Proceedings of the 2002 IEEE Power Engineering Society Winter Meeting, New York, NY, USA, 27–31 January 2002.
8. Liu, M.; Guo, L.; Wang, C.; Zhao, B.; Zhang, X.; Liu, Y. A coordinated operating control strategy for hybrid isolated microgrid including wind power, photovoltaic system, diesel generator and battery storage. *Autom. Electr. Power Syst.* **2012**, *36*, 19–24.
9. Zhang, X.; Yeh, W.; Jiang, Y.; Huang, Y.; Xiao, Y.; Li, L. A case study of control and improved simplified swarm optimization for economic dispatch of a stand-alone modular microgrid. *Energies* **2018**, *11*, 793. [CrossRef]
10. Xiang, Y.; Liu, J.; Liu, Y. Robust energy management of microgrid with uncertain renewable generation and load. *IEEE Trans. Smart Grid* **2016**, *7*, 1034–1043. [CrossRef]
11. Tavakoli, M.; Shokridehaki, F.; Akorede, M.F.; Marzband, M.; Vechiu, I.; Pouresmaeil, E. CVaR-based energy management scheme for optimal resilience and operational cost in commercial building microgrids. *Int. J. Electr. Power Energy Syst.* **2018**, *100*, 1–9. [CrossRef]
12. Huang, C.; Yue, D.; Deng, S.; Xie, J. Optimal scheduling of microgrid with multiple distributed resources using interval optimization. *Energies* **2017**, *10*, 339. [CrossRef]
13. Dutta, S.; Sharma, R. Optimal storage sizing for integrating wind and load forecast uncertainties. In Proceedings of the 2012 IEEE PES Innovative Smart Grid Technologies (ISGT), Washington, DC, USA, 16–20 January 2012.
14. Fan, S.; Ai, Q.; Piao, L. Hierarchical energy management of microgrids including storage and demand response. *Energies* **2018**, *11*, 1111. [CrossRef]
15. Guo, S.; Yuan, Y.; Zhang, X.; Bao, W.; Liu, C.; Cao, Y.; Wang, H. Energy Management Strategy of Isolated Microgrid Based on Multi-time Scale Coordinated Control. *Trans. China Electrotech. Soc.* **2014**, *29*, 122–129.

16. Zhang, C.; Xu, Y.; Dong, Z.Y.; Ma, J. Robust operation of microgrids via two-stage coordinated energy storage and direct load control. *IEEE Trans. Power Syst.* **2017**, *32*, 2858–2868. [CrossRef]

17. Luna, A.C.; Diaz, N.L.; Graells, M.; Vasquez, J.C.; Guerrero, J.M. Mixed-integer-linear-programming-based energy management system for hybrid PV-wind-battery microgrids: Modeling, design and experimental verification. *IEEE Trans. Power Electron.* **2017**, *32*, 2769–2783. [CrossRef]

18. Lee, W.-P.; Choi, J.-Y.; Won, D.-J. Coordination strategy for optimal scheduling of multiple microgrids based on hierarchical system. *Energies* **2017**, *10*, 1336. [CrossRef]

19. Fang, J.; Tang, Y.; Li, H.; Li, X. A battery/ultracapacitor hybrid energy storage system for implementing the power management of virtual synchronous generators. *IEEE Trans. Power Electron.* **2018**, *33*, 2820–2824. [CrossRef]

20. Nguyen-Hong, N.; Nguyen-Duc, H.; Nakanishi, Y. Optimal sizing of energy storage devices in isolated wind-diesel systems considering load growth uncertainty. *IEEE Trans. Ind. Appl.* **2018**, *54*, 1983–1991. [CrossRef]

21. Momayyezan, M.; Abeywardana, D.B.W.; Hredzak, B.; Agelidis, V.G. Integrated reconfigurable configuration for battery/ultracapacitor hybrid energy storage systems. *IEEE Trans. Energy Convers.* **2016**, *31*, 1583–1590. [CrossRef]

22. Ju, C.; Wang, P.; Goel, L.; Xu, Y. A two-layer energy management system for microgrids with hybrid energy storage considering degradation costs. *IEEE Trans. Smart Grid* **2018**. [CrossRef]

23. Xu, G.; Shang, C.; Fan, S.; Hu, X.; Cheng, H. A hierarchical energy scheduling framework of microgrids with hybrid energy storage systems. *IEEE Access* **2018**, *6*, 2472–2483. [CrossRef]

24. Li, J.; Zheng, X.; Ai, X.; Wen, J.; Sun, S.; Li, G. Optimal design of capacity of distributed generation in island standalone microgrid. *Trans. China Electrotech. Soc.* **2016**, *31*, 176–184.

25. Chen, J. Research on Optimal Sizing of Wind/Solar/Battery(/Diesel Generator) Microgrid. Ph.D. Thesis, Tianjin University, Tianjin, China, 2014.

26. Mohamed, E.A.A.E.; Yusuff, M.A.; Wahab, D.A. Application of rainflow cycle counting in the reliability prediction of automotive front corner module system. In Proceedings of the 2009 16th International Conference on Industrial Engineering and Engineering Management, Beijing, China, 21–23 October 2009.

energies

MDPI

Article

A Novel Stability Improvement Strategy for a Multi-Inverter System in a Weak Grid Utilizing Dual-Mode Control

Ming Li [1,*], Xing Zhang [1] and Wei Zhao [2]

[1] School of Electrical Engineering and Automation, Hefei University of Technology, Hefei 230009, China; honglf@ustc.edu.cn
[2] Sungrow Power Supply Co., Ltd., Hefei 230088, China; zhaow@sungrowpower.com
[*] Correspondence: mingjhuu@mail.hfut.edu.cn; Tel.: +86-551-6290-1408

Received: 31 July 2018; Accepted: 14 August 2018; Published: 17 August 2018

Abstract: Due to the increasing penetration of distributed generations (DGS) and non-negligible grid impedance, the instability problem of the multi-inverter system operating in current source mode (CSM) is becoming serious. In this paper, a closed-loop transfer function model of such a multi-inverter system is established, by which it is concluded that output current resonance will occur with the increase in the grid impedance. In order to address this problem, this paper presents a novel dual-mode control scheme of multiple inverters: one inverter operating in CSM will be alternated into voltage source mode (VSM) if the grid impedance is high. It is theoretically proved that the coupling between the inverters and the resonance in the output current can be suppressed effectively with the proposed scheme. Finally, the validity of the proposed theory is demonstrated by extensive simulations and experiments.

Keywords: current source mode (CSM); distributed generations; grid-connected inverter; grid impedance; multi-inverter system; voltage source mode (VSM); weak grid

---

1. Introduction

Constrained by land resources and light resources, large-scale solar photovoltaic power generation and other distributed generations (DGS) are often installed in remote areas, and due to the impact of long transmission lines and transformers, the point of common coupling (PCC) shows weak grid characteristics with non-negligible grid impedance [1–3]. In addition, with the increasing penetration of DGS, it is necessary to connect multiple grid-connected inverters in parallel to the grid in order to increase the power generation efficiency and the scalability of the system [4]. Note that parallel inverters can cause circulating currents and can lead to power semiconductor damage [5–9]. Furthermore, the equivalent grid impedance of a single grid-connected inverter increases, and the grid presents obvious weak grid characteristics [10,11].

At present, there are two main types of grid-connected inverter stability control strategies for weak grids:

The first type is the traditional control strategy of grid-connected inverters, which regulates the active and reactive power injected into the grid by adjusting the *d*-axis and *q*-axis currents based on grid voltage orientation. The phase-locked loop (PLL) is used to observe the voltage phase of the grid. In this control mode, the inverter is equivalent to a current source, which is called current source mode (CSM) in this paper. Moreover, the existing grid-connected inverters use CSM control to achieve maximum power point tracking (MPPT) so as to ensure the maximum efficiency of new energy generation. So, the CSM scheme is a widely used grid-connected control strategy in multi-inverter systems [11]. Generally speaking, when multiple inverters are connected in parallel to the ideal power grid, the output characteristics of each inverter are independent and uncoupled as long as a single inverter can operate stably, so the system can operate stably as well. However, the grid impedance, which cannot be ignored in the weak grid, leads to the coupling between grid-connected inverters, which may lead to the harmonic resonance of the system. Therefore, research on the resonance mechanism and resonance suppression strategy of multi-inverter parallel systems has attracted wide attention. For example, in [12], a passive network model based on the LCL filter is proposed for the weak grid, and the parallel resonance of multiple inverters in DGS is analyzed. In [10], the passive network model is introduced based on the output of the grid-connected inverter, and the equivalent circuit model of the multi-inverter parallel system is constructed. Furthermore, the resonant characteristics of the multi-inverter parallel system with respect to the number of parallel inverters are analyzed. In [13], from the perspective of the digital control of a grid-connected inverter parallel system, the bandwidth and stability of the grid-connected inverter in the weak grid are analyzed. In addition, several works in the literature focus on the stability problem of the multi-inverter system composed of CSM-controlled inverters in the weak grid. For example, in [11], the Norton equivalent circuit model of an inverter using an LCL filter with deadbeat control is established, and the resonant distribution of inverters in parallel and the excitation sources that may cause resonance are analyzed. The virtual parallel resistor is constructed by introducing capacitance harmonic voltage feedback to improve the damping of the system and suppress the resonance. In [14], an equivalent single-inverter system is proposed to analyze the stability of complex multi-inverter systems, and a controller is proposed to improve the stability of the system from the perspective of remolding admittance. In [15], the stability and power quality of the system are studied, and an active damper based on a power electronic converter is proposed. Using the resistive active power filter and voltage resonance compensation control concept, a virtual resistive component damping resonance is constructed at resonant frequency, but some additional devices are needed.

In addition to the above CSM control, the grid-connected inverter stability control strategy of the second type of weak grids is to adjust the active power by adjusting the phase of the output voltage, and the reactive power is adjusted by the change of amplitude of the output voltage vector. The relationship between output power and voltage is similar to that of a synchronous generator system. So, the inverter is equivalent to a voltage source, which is called voltage source mode (VSM) in this paper. To date, several related works in the literature have been studied for the stability of the VSM-controlled inverter in the weak grid. For example, in [16], for the intermittent power generation characteristics of wind power generation, the VSM control in the weak grid is used to allocate the power of the parallel inverter, and the grid-connected inverter stability control in the weak grid is realized. In [17], by adopting the variable pitch control and the coordinated control of the engine speed, the power balance between the wind turbine and the wind engine is realized, and the direct current (DC) side voltage of the inverter is kept constant. However, this paper does not analyze the effect of grid impedance changes on the stability of the VSM-controlled inverter. In [18], by analyzing the transmission characteristics of a wind farm connected with a weak grid, a static synchronous compensator (STATCOM), which is characterized by VSM control, is used to improve the voltage regulation characteristics of wind turbines under a weak grid by voltage and reactive power droop control; however, some additional devices are needed. In [19], by analyzing and comparing the static characteristics of power transmission between CSM and VSM, it is found that VSM is more suitable in an extremely weak grid than CSM, and the grid impedance adaptation dual-mode control strategy in the weak grid is proposed: When the grid impedance is small, the inverter can adopt the traditional CSM, but when the grid becomes weaker, the inverter can be switched from CSM to VSM, which makes the inverter operate stably in a wider range of grid impedance changes. However, [19] only considers the case of a single inverter, and it does not analyze the stability of a single inverter switching from CSM to VSM in a multi-inverter system.

Based on the above literature, it can be found that the stability of multi-inverter system with CSM control has been widely researched. However, the stability analysis of the inverter operating in VSM in a weak grid is rare, and there are no studies in the literature based on the stability of a multi-inverter system with dual-mode control in a weak grid. Since the grid-connected inverter operating in VSM is more stable in an extremely weak grid than CSM [19], a novel stability improvement strategy for a multi-inverter system in a weak grid utilizing dual-mode control is proposed: i.e., one inverter operating in CSM will be alternated into VSM if the grid impedance is high. It is theoretically proved that the coupling between the inverters and the resonance in the output current can be suppressed effectively with the proposed scheme.

This paper is organized as follows. In Section 2, the model of a CSM-only-controlled multi-inverter system in a weak grid is established. Section 3 concluded that output current resonance will occur with the increase in the grid impedance. Section 4 explains the proposed dual-mode control strategy in the weak grid and its stability is analyzed. In Section 5, simulations and experimental results of a multi-inverter system consisting of three 100-kW grid-connected inverters are demonstrated. Section 6 is devoted to the conclusion.

2. Modeling of a Current Source Mode (CSM)-Only-Controlled Multi-Inverter System in a Weak Grid

According to the above introduction, the multi-inverter system often operates under a single-mode control strategy with CSM. Figure 1 shows a typical structure of the three-phase grid-connected inverter operating at CSM, and N such structures are connected in parallel will form a multi-inverter system.

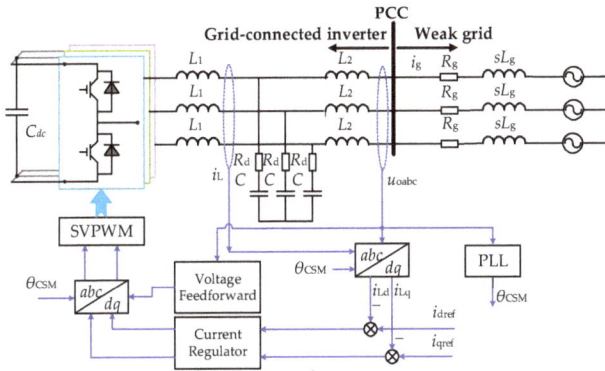

Figure 1. Typical structure of the three-phase grid-connected inverter operating at current source mode (CSM). PLL: phase-locked loop; PCC: point of common coupling; SVPWM: space vector pulse width modulation.

In Figure 1, V_{dc} is the DC voltage, L_1 is the inverter side filter inductor, C is the filter capacitor, R_d is the damping resistor, L_2 is the grid side filter inductor, i_L is the inverter side current, i_g represents the grid current, and e_g is the grid voltage. The i_{dref} and i_{qref} are the reference values of the d axis and the q axis current, respectively. i_{Ld} and i_{Lq} are the d axis and q axis components, respectively, of the current i_L in the synchronous rotating coordinate system. θ_{CSM} is the phase obtained by the PLL according to the PCC voltage u_{oabc}. Z_g is the grid impedance, as shown in Equation (1).

$$Z_g(s) = R_g + sL_g \tag{1}$$

where R_g and L_g represent the resistive component and inductive component of grid impedance, respectively. Since the stability of the inverter in a weak grid is very susceptible to the inductance component, and the resistive component has the effect of enhancing the stability of the inverter [20], this paper takes $R_g = 0$.

According to Figure 1, the control block diagram of an inverter operating in CSM is shown in Figure 2. Since the *d*-axis and *q*-axis controls are the same, the subscripts *d* and *q* are omitted in the following text.

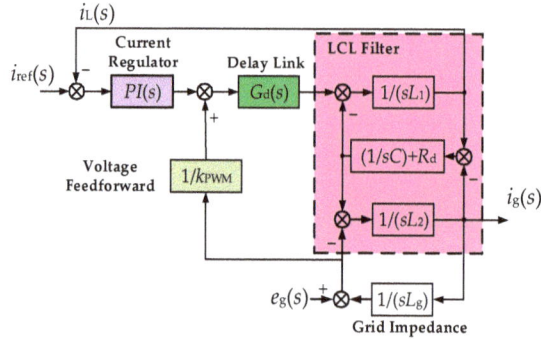

Figure 2. Control block diagram of inverter operating at CSM. *PI*(*s*): proportional-integral controller.

In Figure 2, the $G_d(s)$ represents the influence of the sampler and the delays caused by a digital controller in the *s*-domain. According to [4,21], the transfer function of the delays incurred by the sampler is expressed as $1/T_s$, where T_s is the sampling period. The delays incurred by a digitally controlled system contain the computation delay and the pulse width modulation (PWM) control delay. The expression of $G_d(s)$ is:

$$G_d(s) = \frac{1}{T_s} \cdot e^{-sT_d} \cdot \frac{1 - e^{-sT_s}}{s} k_{PWM} \approx k_{PWM} e^{-1.5sT_s} \tag{2}$$

where T_d is the delay time.

From Figure 2, the transfer function between the input current command $i_{ref}(s)$ and the output current $i_g(s)$ can be obtained:

$$i_g(s) = \frac{i_{ref}(s) \cdot PI(s) \cdot G_d(s) - e_g(s) \cdot \left(\begin{array}{c} s^2 L_1 C/(1 + R_d Cs) + sC/(1 + R_d Cs) \cdot PI(s) \cdot G_d(s) \\ -G_d(s)/k_{PWM} + 1 \end{array} \right)}{\left[\begin{array}{c} s^3 L_1 C/(1 + R_d Cs)(L_2 + L_g) + s^2 C/(1 + R_d Cs)(L_2 + L_g)PI(s) \cdot G_d(s) \\ +s(L_1 + L_2 + L_g) + PI(s) \cdot G_d(s) - sL_g G_d(s)/k_{PWM} \end{array} \right]} \tag{3}$$

According to [19], the inverter operating in CSM is equivalent to the current source, and from Equation (3), the current source equivalent model is shown in Figure 3.

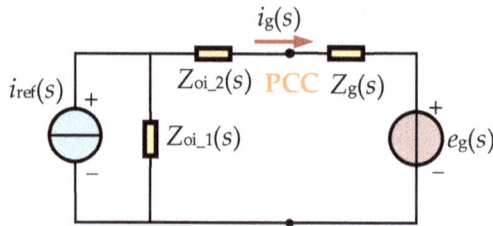

Figure 3. Current source equivalent model of an inverter operating in CSM.

$Z_{oi\_1}(s)$ and $Z_{oi\_2}(s)$ in Figure 3 can be derived from Equation (3):

$$Z_{oi\_1}(s) = \frac{PI(s)G_d(s)}{s^2 L_1 C/(1 + R_d Cs) + sC/(1 + R_d Cs)PI(s)G_d(s) + 1 - G_d(s)} \tag{4}$$

$$Z_{oi\_2}(s) = \frac{sL_1}{s^2 L_1 C/(1 + R_d Cs) + sC/(1 + R_d Cs)PI(s)G_d(s) + 1 - G_d(s)} + sL_2 \tag{5}$$

Generally, photovoltaic and wind power plants employ the same type of the inverter with identical hardware parameters and control algorithms, so the equivalent models of the inverter circuits are consistent. According to Figure 3, an equivalent model of a multi-inverter system with CSM-only control is available, as shown in Figure 4. N represents the number of inverters; for a multi-inverter system, N is larger than 2. i represents any inverter in this system, and $1 \leq i \leq N$.

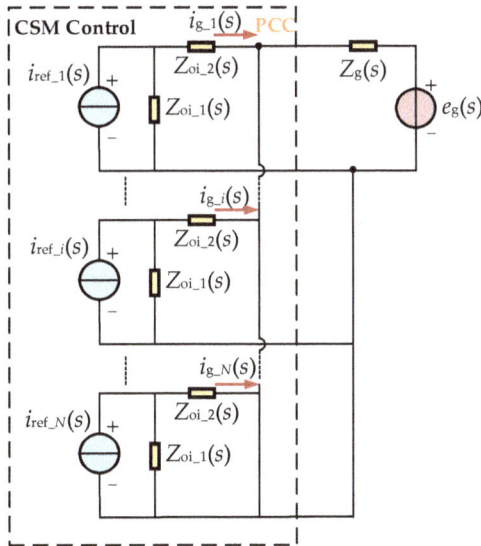

Figure 4. Equivalent model of a multi-inverter system with CSM-only control.

From Figure 4, we can get:

$$\begin{bmatrix} i_{g\_1}(s) \\ \vdots \\ i_{g\_i}(s) \\ \vdots \\ i_{g\_N}(s) \end{bmatrix} = \begin{bmatrix} G_{CSM\_P}(s) & \cdots & G_{CSM\_N}(s) & \cdots & G_{CSM\_N}(s) \\ \vdots & \ddots & \vdots & & \vdots \\ G_{CSM\_N}(s) & \cdots & G_{CSM\_P}(s) & \cdots & G_{CSM\_N}(s) \\ \vdots & & \vdots & \ddots & \vdots \\ G_{CSM\_N}(s) & \cdots & G_{CSM\_N}(s) & \cdots & G_{CSM\_P}(s) \end{bmatrix} \begin{bmatrix} i_{ref\_1}(s) \\ \vdots \\ i_{ref\_i}(s) \\ \vdots \\ i_{ref\_N}(s) \end{bmatrix} + G_{CCM\_E}(s)E(s) \tag{6}$$

where

$$G_{CCM\_P}(s) = \frac{Z_{oi\_1}(s)}{Z_{oi\_1}(s) + Z_{oi\_2}(s) + Z_g(s) // \left(\frac{Z_{oi\_1}(s) + Z_{oi\_2}(s)}{N-1} \right)} \tag{7}$$

$$G_{CCM\_N}(s) = \frac{Z_{oi\_1}(s)\frac{Z_g(s)}{(N-1)Z_g(s) + Z_{oi\_1}(s) + Z_{oi\_2}(s)}}{Z_{oi\_1}(s) + Z_{oi\_2}(s) + Z_g(s) // \left(\frac{Z_{oi\_1}(s) + Z_{oi\_2}(s)}{N-1} \right)} \tag{8}$$

$$G_{CCM\_E}(s) = \frac{1}{NZ_g(s) + Z_{oi\_1}(s) + Z_{oi\_2}(s)} \tag{9}$$

In Equations (7) and (8), "//" represents the parallel operation, i.e., $x//y = xy/(x+y)$.

3. Stability Analysis of a CSM-Only-Controlled Multi-Inverter System in a Weak Grid

According to Equation (6), due to the existence of grid impedance, multiple coupling transfer functions exist in multi-inverter systems, which affects the system stability. To illustrate the effect of grid impedance on this multi-inverter system, the main parameters of the inverter operating in CSM are given in Table 1.

Table 1. The main parameters of the inverter operating in current source mode (CSM).

Hardware Parameters of Inverter	
Parameters	Value
Inverter capacity	100 kW
Inverter voltage and frequency	380 V, 50 Hz
DC voltage	$V_{dc} = 600$ V
LCL filter	$L_1 = 0.56$ mH, $C = 90$ μF, $R_d = 0.5$ Ω, $L_2 = 0.09$ mH
Switching frequency	10 kHz
Reference value of grid impedance	$Z_g(s) = 0.25$ mH = 1 p.u.
Controller Parameters of Inverter Operating in CSM	
Parameters	Value
Current PI regulator	$PI(s) = 1.5 + 500/s$
PWM gain	$k_{PWM} = 1$
Sampling time	$T_s = 1$ ms
Delay function	$G_d(s) = k_{PWM} \cdot e^{-s*1.5*Ts} = e^{-s*1.5*0.001}$

PWM: pulse width modulation.

In this section, the multi-inverter system takes the number of units $N = 3$ as an example and takes the output current of the first inverter ($i = 1$) as the research object, and the characteristics of the transfer functions $G_{CSM\_P}(s)$, $G_{CSM\_N}(s)$, and $G_{CSM\_E}(s)$ are studied, respectively.

$G_{CSM\_P}(s)$ represents the closed-loop transfer function of the reference current to the output current of the CSM-controlled inverters. The Bode diagram and pole-zero map of $G_{CSM\_P}(s)$ with the grid impedance $Z_g(s)$ changes from 1 p.u. to 6 p.u. are shown in Figures 5a and 5b, respectively. From Figure 5a, when the grid impedance increases, the amplitude of the resonance peak gets larger, and its frequency is gradually moving toward the low-frequency direction. When $Z_g(s) = 6$ p.u., the amplitude of the resonance peak is nearly 20 dB, which means that the current harmonics here will be amplified about 10 times, and the output current will resonate. To further illustrate the stability issue, the distribution of poles near the imaginary axis is given in Figure 5b. With a $Z_g(s)$ increase from 1 p.u. to 6 p.u., the poles gradually cross the imaginary axis, so the inverter will operate in an unstable fashion.

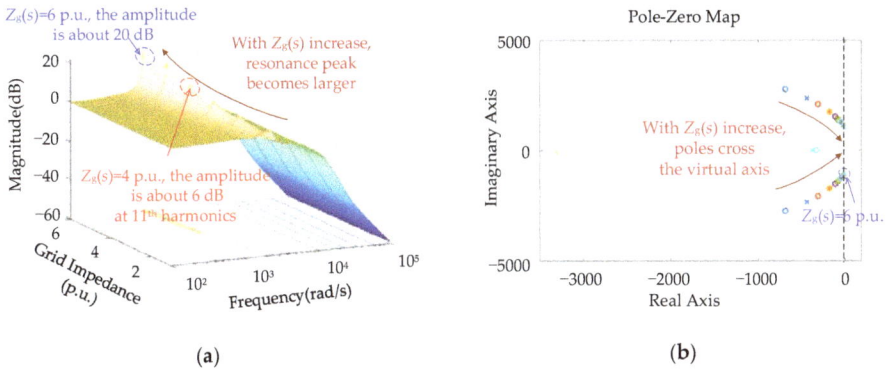

(a)   (b)

Figure 5. Grid impedance changes in the multi-inverter system with CSM-only control: (**a**) Bode diagram of $G_{CSM\_P}(s)$ with the grid impedance $Z_g(s)$ changes from 1 p.u. to 6 p.u.; (**b**) pole-zero map of $G_{CSM\_P}(s)$ with the grid impedance $Z_g(s)$ changes from 1 p.u. to 6 p.u.

$G_{CSM\_N}(s)$ represents the interaction between the CSM-controlled inverters. The Bode diagram and pole-zero map of $G_{CSM\_N}(s)$ with the grid impedance $Z_g(s)$ changes from 1 p.u. to 6 p.u. are shown in Figures 6a and 6b, respectively. From Figure 6a, when the grid impedance increases, the amplitude of the resonance peak appears and becomes larger. When $Z_g(s) = 6$ p.u., the amplitude of resonance peak is nearly 20 dB at 1000 rad/s, which means that the current harmonics of one inverter will be amplified in another inverter by about 10 times. Therefore, with the increase of grid impedance, the coupling between the inverters increases, and the harmonics of the other inverters will be amplified and become resonant at the output current of other inverters. To further illustrate the stability issue, the distribution of poles near the imaginary axis is shown in Figure 6b. With the $Z_g(s)$ increase from 1 p.u. to 6 p.u., the poles gradually cross the imaginary axis, so the inverter will operate in an unstable fashion.

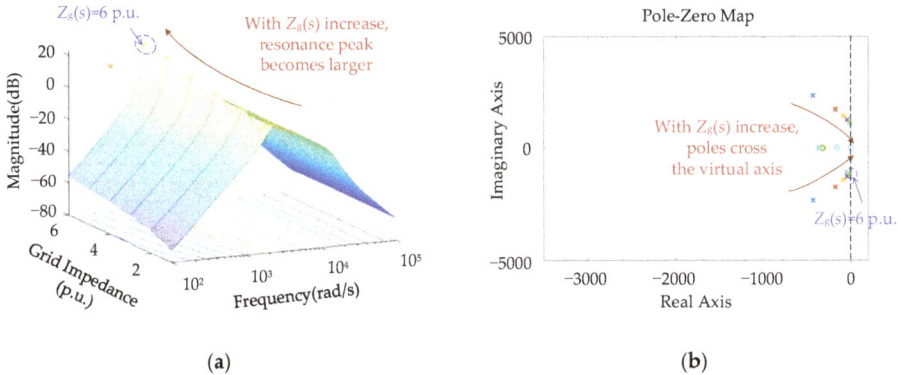

(a)   (b)

Figure 6. Grid impedance changes in the multi-inverter system with CSM-only control: (**a**) Bode diagram of $G_{CSM\_N}(s)$ with the grid impedance $Z_g(s)$ changes from 1 p.u. to 6 p.u.; (**b**) pole-zero map of $G_{CSM\_N}(s)$ with the grid impedance $Z_g(s)$ changes from 1 p.u. to 6 p.u.

$G_{CSM\_E}(s)$ represents the impact between the CSM-controlled inverters and the grid. The Bode diagram and pole-zero map of $G_{CSM\_E}(s)$ with the grid impedance $Z_g(s)$ changes from 1 p.u. to 6 p.u. are shown in Figures 7a and 7b, respectively. Comparing Figures 6 and 7, it can be seen that the Bode diagram and the pole-zero map of the transfer function are very similar, so there is a similar conclusion: with the grid impedance increases, the degree of coupling between the grid and the inverter increases, and the background harmonics of the weak grid will be amplified and become resonant at the output current of the inverter.

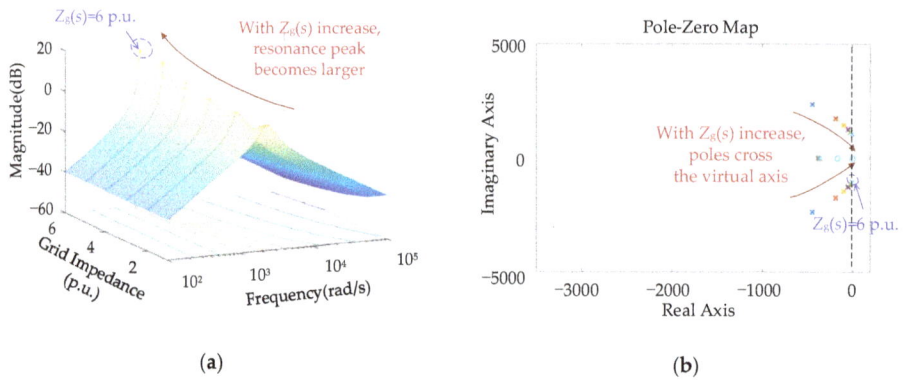

(a)

(b)

Figure 7. Grid impedance changes in the multi-inverter system with CSM-only control: (a) Bode diagram of $G_{CSM\_E}(s)$ with the grid impedance $Z_g(s)$ changes from 1 p.u. to 6 p.u.; (b) pole-zero map of $G_{CSM\_E}(s)$ with the grid impedance $Z_g(s)$ changes from 1 p.u. to 6 p.u.

In summary, by constructing the closed-loop transfer function model of the multi-inverter system with CSM-only control, it is found that as the grid impedance increases, a resonance peak appears and the degree of coupling between the inverters and the grid increases, and the inverter is prone to output current resonance.

4. Proposed Dual-Mode Control Strategy in the Weak Grid

Previous research [19] points out that the grid-connected inverter operating in VSM is more stable in a weak grid than CSM. Due to the object of analysis in this study being only a single inverter, if the inverters in a multi-inverter system are equipped with such a dual-mode adaptive switching operation control strategy which could adaptively switch the operating mode between CSM and VSM, then a novel control strategy for multi-inverter system stability based on dual mode grid impedance adaptation in the weak grid is proposed: One inverter operating in CSM in the system could switch to VSM control based on the grid impedance identification, and then the multi-inverter system will run under dual-mode control and operate stably.

In order to validate this proposed control strategy, similar to the above analysis in Section 2, the model of a multi-inverter system with dual modes is established.

Figure 8 shows a typical structure of the three-phase grid-connected inverter operating in VSM. The main circuit of the inverter is the same as that of Figure 1. u_{dref} and u_{qref} are the d-axis and q-axis components of the reference voltage loop, respectively. θ_{VSM} is the phase of the power angle which can be set according to the desired output power. The u_d and u_q are the d-axis and q-axis components of the capacitor voltage u_{Cabc}, respectively. Differently from CSM control, the inverter operating in VSM is equivalent to a voltage source.

According to Figure 8, the control block diagram of the inverter operating in VSM is shown in Figure 9.

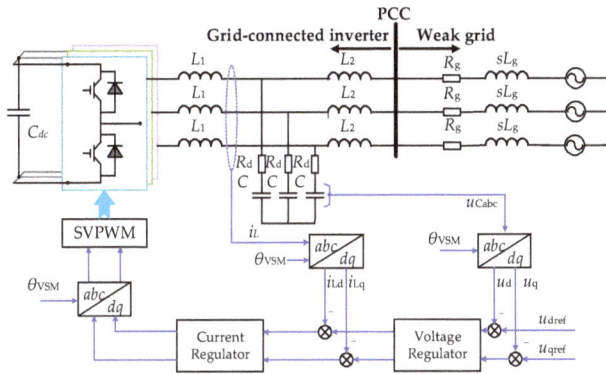

Figure 8. Typical structure of the three-phase grid-connected inverter operating in voltage source mode (VSM).

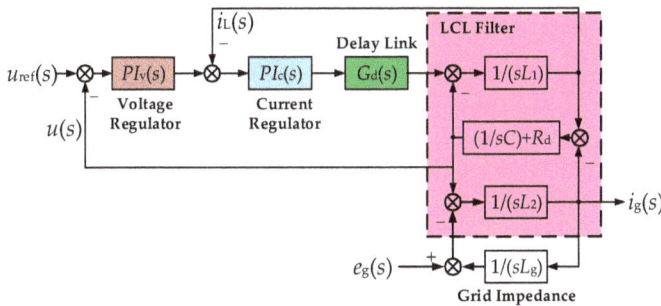

Figure 9. Control block diagram of an inverter operating in VSM.

In Figure 9, the $PI_v(s)$ represents the voltage outer loop proportional-integral (PI) regulator, and $PI_c(s)$ is the current inner loop proportional regulator. The expressions of $PI_v(s)$ and $PI_c(s)$ are shown as follows:

$$PI_v(s) = K_{vP} + (K_{vI}/s) \tag{10}$$

$$PI_c(s) = K_{iP} \tag{11}$$

where K_{vP} and K_{vI} are proportional and integral coefficients of the $PI_v(s)$, respectively. K_{iP} is the proportional coefficient of the $PI_c(s)$.

According to Figure 9, the transfer function between the input voltage command $u_{ref}(s)$ and the output voltage $u(s)$ can be obtained [22]:

$$u(s) = u_{ref}(s) \frac{PI_v(s)}{PI_v(s) \cdot PI_c(s)G_d(s) + sC/(1 + R_dCs)(sL_1 + PI_c(s)G_d(s)) + 1}$$
$$-i_g(s) \left(\frac{sL_1 + PI_c(s)G_d(s)}{PI_v(s) \cdot PI_c(s)G_d(s) + sC/(1 + R_dCs)(sL_1 + PI_c(s)G_d(s)) + 1} + sL_2 \right) \tag{12}$$

From Equation (12), the voltage source equivalent model is shown in Figure 10; $G_v(s)$ and $Z_{ov}(s)$ in Figure 10 can be derived from Equation (12):

$$G_v(s) = \frac{PI_v(s)}{PI_v(s) \cdot PI_c(s)G_d(s) + sC/(1 + R_dCs)(sL_1 + PI_c(s)G_d(s)) + 1} \tag{13}$$

$$Z_{ov}(s) = \frac{sL_1 + PI_c(s)G_d(s)}{PI_v(s) \cdot PI_c(s)G_d(s) + sC/(1 + R_dCs)(sL_1 + PI_c(s)G_d(s)) + 1} + sL_2 \tag{14}$$

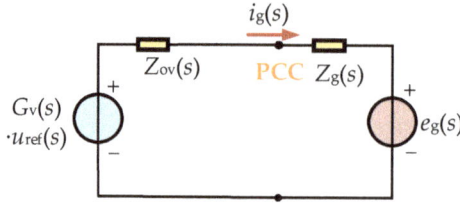

Figure 10. Voltage source equivalent model of an inverter operating in VSM.

4.1. Modeling of the Multi-Inverter System with Dual Modes

According to Figures 3 and 10, when the multi-inverter system contains CSM- and VSM-controlled inverters at the same time, the equivalent model of a multi-inverter system with dual modes is shown in Figure 11.

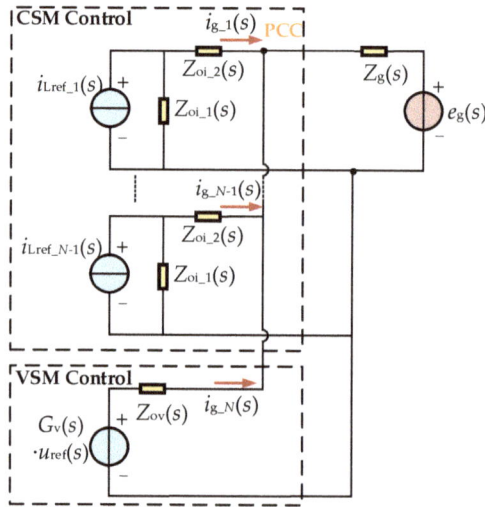

Figure 11. Equivalent model of a multi-inverter system with dual modes.

From Figure 11, we can get:

$$\begin{bmatrix} i_{g\_1}(s) \\ \vdots \\ i_{g\_i}(s) \\ \vdots \\ i_{g\_N-1}(s) \end{bmatrix} = \begin{bmatrix} G_{VSM\_P}(s) & \cdots & G_{VSM\_N}(s) & \cdots & G_{VSM\_N}(s) \\ \vdots & \ddots & \vdots & & \vdots \\ G_{VSM\_N}(s) & \cdots & G_{VSM\_P}(s) & \cdots & G_{VSM\_N}(s) \\ \vdots & & \vdots & \ddots & \vdots \\ G_{VSM\_N}(s) & \cdots & G_{VSM\_N}(s) & \cdots & G_{VSM\_P}(s) \end{bmatrix} \begin{bmatrix} i_{ref\_1}(s) \\ \vdots \\ i_{ref\_i}(s) \\ \vdots \\ i_{ref\_N-1}(s) \end{bmatrix} \tag{15}$$
$$+ G_{VSM\_E}(s)e_g(s) + G_{VSM\_U}(s)u_{ref}(s)$$

where

$$G_{VCM\_P}(s) = \frac{Z_{oi\_1}(s)}{Z_{oi\_1}(s) + Z_{oi\_2}(s) + Z_g(s)//\left(\frac{Z_{oi\_1}(s)+Z_{oi\_2}(s)}{N-2}\right)//Z_{ov}(s)} \tag{16}$$

$$G_{VCM\_N}(s) = \frac{-Z_{oi\_1}(s)\left(\frac{Z_g(s)}{Z_{oi\_1}(s)+Z_{oi\_2}(s)+(N-2)Z_g(s)}//Z_{ov}(s)\right)}{Z_{oi\_1}(s) + Z_{oi\_2}(s) + Z_g(s)//\left(\frac{Z_{oi\_1}(s)+Z_{oi\_2}(s)}{N-2}\right)//Z_{ov}(s)} \tag{17}$$

$$G_{VCM\_E}(s) = -\frac{Z_{ov}(s)}{(Z_g(s) + Z_{ov}(s))(Z_{oi\_1}(s) + Z_{oi\_2}(s)) + (N-1)Z_{ov}(s)Z_g(s)} \tag{18}$$

$$G_{VCM\_U}(s) = -\frac{Z_g(s)G_v(s)}{(Z_g(s) + Z_{ov}(s))(Z_{oi\_1}(s) + Z_{oi\_2}(s)) + (N-1)Z_{ov}(s)Z_g(s)} \tag{19}$$

In Equations (16) and (17), "$//$" represents the parallel operation, i.e., $x//y = xy/(x+y)$.

4.2. Stability Analysis of the Multi-Inverter System with Dual Modes

In Section 2, the stability of the multi-inverter system with CSM-only control is analyzed; a similar analysis method will be used in this section to verify that the proposed dual-mode multi-inverter system can effectively improve system stability and reduce coupling between inverters in a weak grid.

The main parameters of the inverter operating in VSM are given in Table 2. It is worth mentioning that the hardware parameters of the VSM-controlled inverter are consistent with the CSM. Therefore, Table 2 only gives the corresponding controller parameters.

Table 2. The main parameters of the inverter operating in voltage source mode (VSM).

Controller Parameters of Inverter Operating in VSM	
Parameters	**Value**
Voltage PI regulator	$PI_v(s) = 1500 + 1/s$
Current proportional regulator	$PI_c(s) = 10$
PWM gain	$k_{PWM} = 1$
Sampling time	$T_s = 1$ ms
Delay function	$G_d(s) = k_{PWM} \cdot e^{-s*1.5*Ts} = e^{-s*1.5*0.001}$

In this section, the multi-inverter system takes the number of units $N = 3$ as an example. This system consists of two CSM-controlled inverters and one VSM-controlled inverter to form a dual-mode hybrid system. Moreover, according to reference [19], grid-connected inverters running in VSM can operate stably in a weak grid with high grid impedance, while grid-connected inverters in CSM are prone to output current resonance. Therefore, the stability of this dual-mode multi-inverter system is studied from the point of view of a CSM-controlled inverter in this paper.

In order to correspond to the analysis of Section 2, the output current of the first CSM-controlled inverter ($i = 1$) in the multi-inverter system is also taken as the research object, and the characteristics of the transfer functions $G_{VSM\_P}(s)$, $G_{VSM\_N}(s)$, $G_{VSM\_E}(s)$, and $G_{VSM\_U}(s)$ are studied, respectively.

Similar to the multi-inverter system with CSM-only control, $G_{VSM\_P}(s)$ represents the closed-loop transfer function of the reference current to the output current of the CSM-controlled inverters in the dual-mode controlled multi-inverter system. The Bode diagram and pole-zero map of $G_{VSM\_P}(s)$ with the grid impedance $Z_g(s)$ changes from 1 p.u. to 6 p.u. are shown in Figures 12a and 12b, respectively. Differently from Figure 5, with the increase of grid impedance, the output current of the grid-connected inverter controlled by the current source has no resonance peak, and the pole is far from the virtual axis. It shows that the inverter can operate stably at this time.

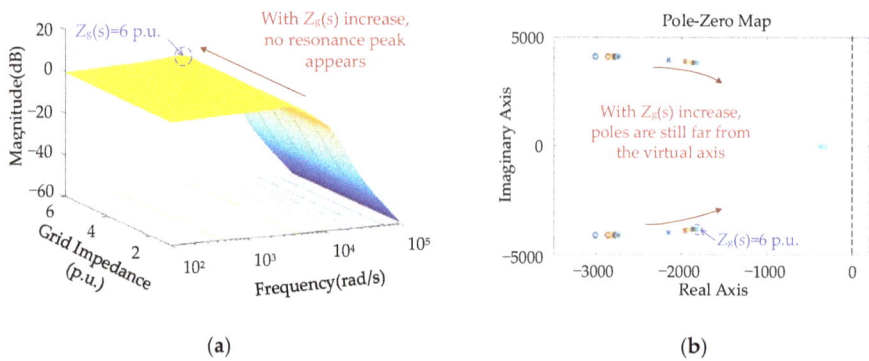

Figure 12. The multi-inverter system with dual-mode control: (**a**) Bode diagram of $G_{VSM\_P}(s)$ with the grid impedance $Z_g(s)$ changes from 1 p.u. to 6 p.u.; (**b**) pole-zero map of $G_{VSM\_P}(s)$ with the grid impedance $Z_g(s)$ changes from 1 p.u. to 6 p.u.

Similarly, $G_{VSM\_N}(s)$ represents the interaction between the CSM-controlled inverters. The Bode diagram and pole-zero map of $G_{VSM\_N}(s)$ with the grid impedance $Z_g(s)$ changes from 1 p.u. to 6 p.u. are shown in Figures 13a and 13b, respectively. With the increase of grid impedance, the output current of the grid-connected inverter controlled by CSM has no resonance peak and the poles are still far from the virtual axis. Therefore, with the grid impedance increases, the degree of coupling between the inverters does not change significantly.

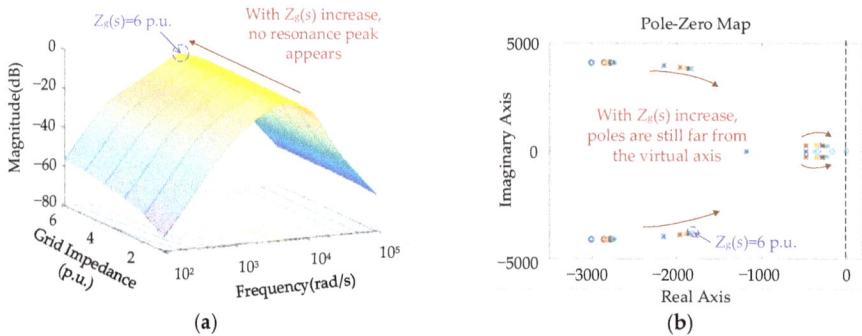

Figure 13. The multi-inverter system with dual-mode control: (**a**) Bode diagram of $G_{VSM\_N}(s)$ with the grid impedance $Z_g(s)$ changes from 1 p.u. to 6 p.u.; (**b**) pole-zero map of $G_{VSM\_N}(s)$ with the grid impedance $Z_g(s)$ changes from 1 p.u. to 6 p.u.

Similarly, $G_{VSM\_E}(s)$ represents the coupling degree between the CSM-controlled inverters and grid. The Bode diagram and pole-zero map of $G_{VSM\_E}(s)$ with the grid impedance $Z_g(s)$ changes from 1 p.u. to 6 p.u. are shown in Figures 14a and 14b, respectively. With the increase of grid impedance, the output current of the grid-connected inverter controlled by CSM has no resonance peak and the poles are still far from the virtual axis. Therefore, as the grid impedance increases, the degree of coupling between the inverters and the grid does not have apparent changes.

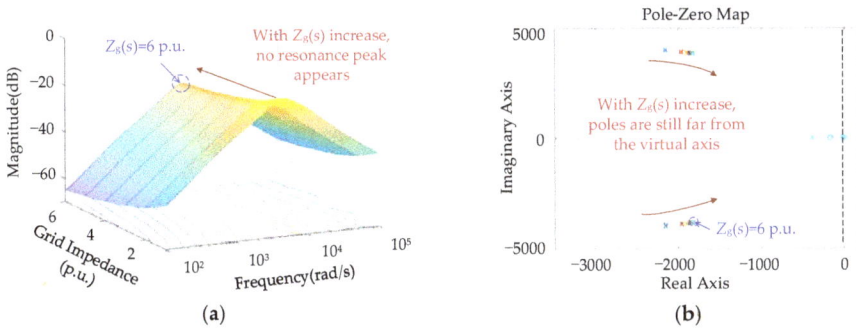

Figure 14. The multi-inverter system with dual-mode control: (**a**) Bode diagram of $G_{VSM\_E}(s)$ with the grid impedance $Z_g(s)$ changes from 1 p.u. to 6 p.u.; (**b**) pole-zero map of $G_{VSM\_E}(s)$ with the grid impedance $Z_g(s)$ changes from 1 p.u. to 6 p.u.

It is worth noting that a new transfer function, $G_{VSM\_U}(s)$, appears when the multi-inverter system is operating under the dual-mode control. $G_{VSM\_U}(s)$ represents the degree of coupling between the CSM-controlled inverters and the VSM-controlled inverter. The Bode diagram and pole-zero map of $G_{VSM\_U}(s)$ with the grid impedance $Z_g(s)$ changes from 1 p.u. to 6 p.u. are shown in Figures 15a and 15b, respectively. It can be seen from Figure 15 that there is no resonance peak on the Bode diagram when the grid impedance increases, and the poles are still far from the virtual axis. Therefore, in this kind of dual-mode multi-inverter system, the coupling between CSM- and VSM-controlled inverters is not increased.

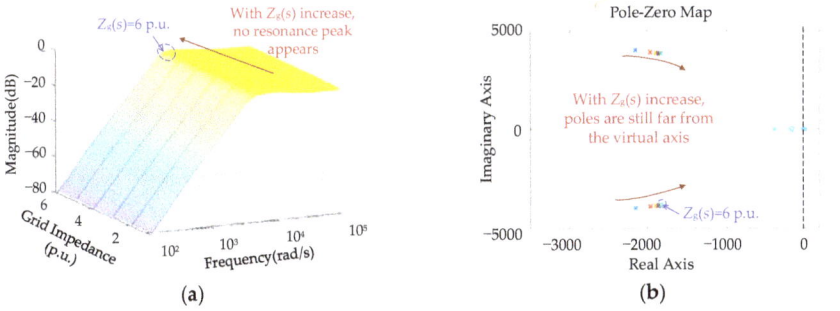

Figure 15. The multi-inverter system with dual-mode control: (**a**) Bode diagram of $G_{VSM\_U}(s)$ with the grid impedance $Z_g(s)$ changes from 1 p.u. to 6 p.u.; (**b**) pole-zero map of $G_{VSM\_U}(s)$ with the grid impedance $Z_g(s)$ changes from 1 p.u. to 6 p.u.

In summary, since the grid-connected inverter operating in voltage source mode (VSM) is more stable in an extremely weak grid than CSM [19], a novel stability improvement strategy of the multi-inverter system in a weak grid utilizing dual-mode control is proposed: one inverter operating in CSM will be alternated into VSM if the grid impedance is high. It is theoretically proved that the coupling between the inverters and the resonance in the output current can be suppressed effectively with the proposed scheme.

5. Simulation and Experimental Results

The proposed dual-mode control strategy in a weak grid is confirmed through Matlab/Simulink and an experiment. Figure 16 shows the multi-inverter system with three parallel grid-connected inverters in a weak grid. The system parameters of the simulation as well as the experiments are almost the same and are enumerated in Tables 1 and 2.

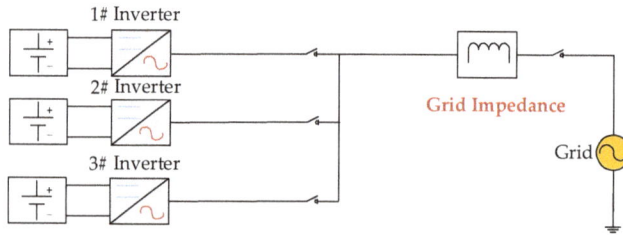

Figure 16. The multi-inverter system with three parallel grid-connected inverters in a weak grid.

In Figure 16, the grid impedance is set to $Z_g(s) = 4$ p.u., and the 3# inverter is equipped with the grid impedance dual-mode strategy, so the inverter can adaptively switch between CSM and VSM according to the grid impedance. Moreover, it can be seen from Figure 5 that when the grid impedance is $Z_g(s) = 3$ p.u., the amplitude of the resonance peak is about 6 dB; that is, the harmonic will be amplified about two times. Therefore, for the specific parameters of the inverter in this paper, as shown in Tables 1 and 2, the grid impedance $Z_g(s) = 3$ p.u. is set to the boundary value of CSM and VSM. Based on this value, Figure 17 shows the flow chart of the switching process between CSM and VSM. Notably, there are many kinds of grid impedance identification schemes [23–27], and the method of single harmonic injection [23] is used in this paper. Since the grid impedance identification scheme is not the core work of this paper, it will not be discussed in this paper.

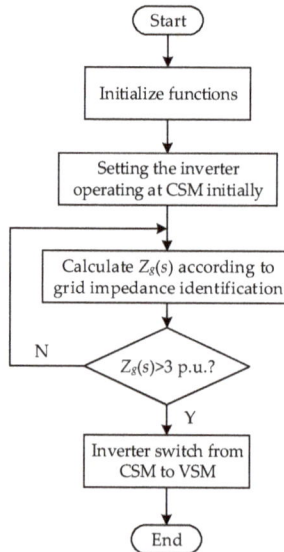

Figure 17. Flow chart of the switching process between CSM and VSM.

In order to avoid chattering between the two modes, a hysteresis switching process as shown in Figure 18 is used in the actual simulation and experiment. As can be seen from Figure 18, the hysteresis center is set to the boundary value $Z_g(s) = 3$ p.u., and the hysteresis width is set to δ, where $\delta = 5\% \times 3$ p.u. = 0.15 p.u.. At this point, the mode switching process of the grid-connected inverter is described as follows: When the grid becomes stronger to weaker, i.e., the grid impedance $Z_g(s) > (3 + \delta)$ p.u., the inverter should switch from CSM to VSM; and when the grid becomes weaker to stronger, i.e., the grid impedance $Z_g(s) < (3 - \delta)$ p.u., the inverter should switch from VSM to CSM.

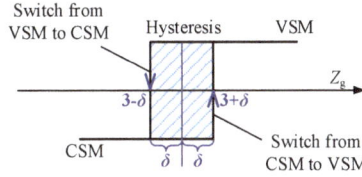

Figure 18. The hysteresis switching process between CSM and VSM control.

5.1. Simulation Verification

The simulation conditions are described as follows:

Figure 19a shows the simulation waveform of the A-phase grid current i_{ga} of the 1# inverter when the multi-inverter system is converted from the original CSM-only control at 0.8 s to the dual-mode control system.

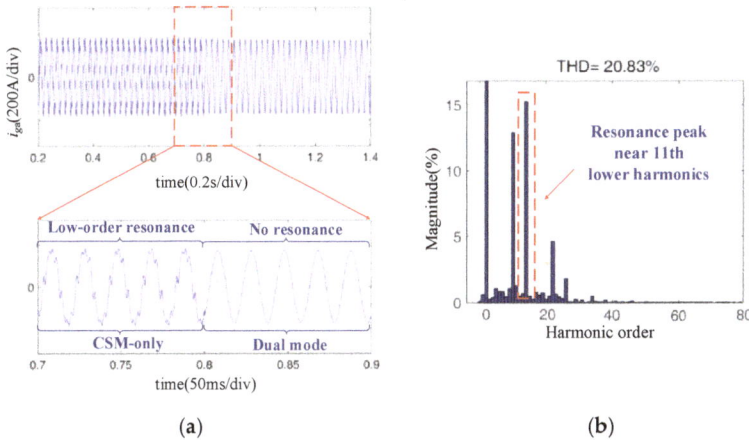

(a)   (b)

Figure 19. (**a**) Simulation waveform of the A-phase grid current i_{ga} of the 1# inverter when the multi-inverter system is converted from the original CSM-only control to the dual-mode control system; (**b**) spectrum of the A-phase grid current i_{ga} before 0.8 s. THD: total harmonic distortion.

Before 0.8 s, all three inverters in the system are running in CSM. It can be found that the current waveform has obvious low-order resonance before 0.8 s, and Figure 19b shows the spectrum of the A-phase grid current i_{ga} before 0.8 s. From Figure 19b, the resonance peak is near the 11th harmonics, and this is consistent with the conclusion that when the grid impedance in Figure 5a is $Z_g(s) = 4$ p.u., and the resonant peak frequency is around the 11th harmonics.

After 0.8 s, due to the 3# inverter in the system having a dual-mode adaptive control algorithm based on the grid impedance added, and since $Z_g(s) = 4$ p.u. > 3 p.u., it will adaptively switch from CSM control to VSM at 0.8 s. After 0.8 s, it is found that the output current resonance phenomenon disappears remarkably and the output current quality is high.

In summary, it can be seen that the dual-mode multi-inverter control strategy proposed in this paper can effectively suppress the output current resonance phenomenon caused by the grid impedance in the weak grid and improve the stability of the grid-connected inverter system.

5.2. Experimental Verification

Experiments were also carried out on an experimental platform consisting of three 100-kW grid-connected inverters to verify the proposed control scheme. The control circuit is implemented on a DSP chip TMS320F28335, which implements the proposed control strategies, as described in the previous sections. In the experiments, a three-phase programmable rectifier is used to represent the DC source. The grid impedance is achieved by a series of practical reactors in a reactor cabinet, and the corresponding grid inductance is 0.25×4 mH (that is, $Z_g(s) = 4$ p.u.). The experimental platform is depicted in Figure 20.

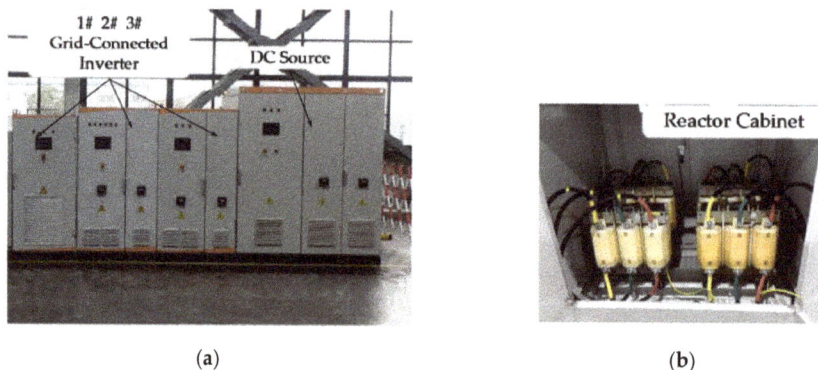

(a)   (b)

Figure 20. Experimental platform: (**a**) Three grid-connected inverter and DC source; (**b**) reactor cabinet to simulate the grid impedance.

Corresponding to the simulation, the experimental conditions are described as follows:

Figure 21a shows the experimental waveform of the A-phase grid current i_{ga} of the 1# inverter when the multi-inverter system is converted from the original CSM-only control at time t_0 to the dual-mode control system.

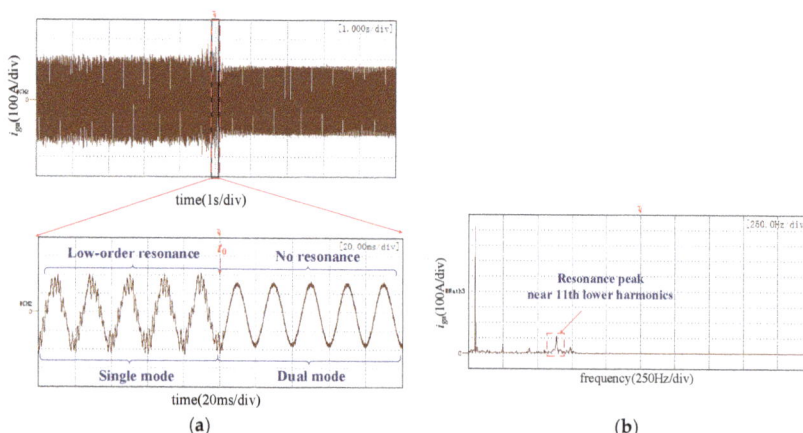

(a)   (b)

Figure 21. (**a**) Experimental waveform of the A-phase grid current i_{ga} of the 1# inverter when the multi-inverter system is converted from the original CSM-only control to the dual-mode control system; (**b**) spectrum of the A-phase grid current i_{ga} before t_0.

Before t_0, all three inverters in the system are running in CSM. Similarly to the simulation results, the current waveform has obvious low-order resonance before 0.8 s, and Figure 21b shows the spectrum of the A-phase grid current i_{ga} before t_0. From Figure 21b, the resonance peak is near the 11th harmonics, and this is consistent with the conclusion that when the grid impedance in Figure 5a is $Z_g(s) = 4$ p.u., the resonant peak frequency is around the 11th harmonics.

After t_0, due to the 3# inverter in the system having a dual-mode adaptive control algorithm based on the grid impedance added, and since $Z_g(s) = 4$ p.u. > 3 p.u., it will adaptively switch from CSM control to VSM at t_0. After t_0, it is found that the output current resonance phenomenon disappears remarkably and the output current quality is high.

It can be concluded from Figure 19 to Figure 21 that the experimental results are in good agreement with the simulation results, and thus the superiority of the proposed control strategy is further verified.

6. Conclusions

In this paper, a closed-loop transfer function model of the multi-inverter system operating in current source mode (CSM) is established, by which it is concluded that output current resonance will occur with the increase in the grid impedance. In order to address this problem, this paper presents a novel dual-mode control scheme of multiple inverters; i.e., one inverter operating in CSM will be alternated into voltage source mode (VSM) if the grid impedance is high. It is theoretically proved that the coupling between the inverters and the resonance in the output current can be suppressed effectively with the proposed scheme. Finally, the validity of the proposed theory is demonstrated by extensive simulations and experiments.

Author Contributions: M.L. designed the algorithm, performed the experiments, and conduced simulations as the first author. X.Z. and W.Z. handled the project and contributed materials and tools. All the authors discussed the simulation and experimental results and approved the publication.

Funding: This work was funded by the National Key Research and Development Program of China (NO. 2016YFB0900300) and the National Natural Science Foundation of China (NO. 51677049).

Acknowledgments: The authors declare no acknowledgments.

Conflicts of Interest: The authors declare no conflict of interest.

Nomenclatures

List of Abbreviations

DGS	distributed generations
PCC	common coupling point
PLL	phase-locked loop
CSM	current source mode
VSM	voltage source mode
MPPT	maximum power point tracking

List of Symbols

V_{dc}	DC voltage
L_1	inverter side filter inductor
C	filter capacitor
R_d	damping resistor
L_2	grid side filter inductor
i_L	inverter side current
i_g	grid current
i_{dref}	reference values of the d axis current
i_{qref}	reference values of the q axis current
i_{Ld}	d axis component of the current i_L

i_{Lq}	q axis component of the current i_L
θ_{CSM}	phase obtained by the PLL according to the PCC voltage
U_{oabc}	PCC voltage
Z_g	grid impedance
R_g	resistive component of grid impedance
L_g	inductive component of grid impedance
N	the number of inverters, for a multi-inverter system
i	any inverter in a multi-inverter system
k_{PWM}	PWM gain
PI	current PI regulator
G_d	delay function
u_{dref}	d axis components of the reference voltage loop
u_{qref}	q axis components of the reference voltage loop
u_{qref}	q axis components of the reference voltage loop
u_d	d axis components of the capacitor voltage
u_q	q axis components of the capacitor voltage
u_{Cabc}	capacitor voltage
PI_v	voltage outer loop PI regulator
PI_c	current inner loop proportional regulator
G_{CSM_P}	closed-loop transfer function of the reference current to the output current of the CSM-controlled inverters
G_{CSM_N}	closed-loop transfer function representing the interaction between the CSM-controlled inverters
G_{CSM_E}	closed-loop transfer function representing the impact between the CSM-controlled inverters and the grid
G_{VSM_P}	closed-loop transfer function of the reference current to the output current of the CSM-controlled inverters in a dual-mode-controlled multi-inverter system
G_{VSM_N}	closed-loop transfer function representing the interaction between the CSM-controlled inverters in a dual-mode-controlled multi-inverter system
G_{VSM_E}	closed-loop transfer function representing the impact between the CSM-controlled inverters and the grid in a dual-mode-controlled multi-inverter system
G_{VSM_U}	closed-loop transfer function representing the coupling degree between the CSM-controlled inverters and the grid in a dual-mode-controlled multi-inverter system

References

1. Ben Said-Romdhane, M.; Naouar, M.W.; Belkhodja, I.S.; Monmasson, E. An improved LCL filter design in order to ensure stability without damping and despite large grid impedance variations. *Energies* **2017**, *10*, 336. [CrossRef]
2. Blaabjerg, F.; Chen, Z.; Kjaer, S.B. Power electronics as efficient interface in dispersed power generation systems. *IEEE Trans. Power Electron.* **2004**, *19*, 1184–1194. [CrossRef]
3. Davari, M.; Mohamed, Y.A.R.I. Robust vector control of a very weak-grid-connected voltage-source converter considering the phase-locked loop dynamics. *IEEE Trans. Power Electron.* **2017**, *32*, 977–994. [CrossRef]
4. Agorreta, J.L.; Borrega, M.; López, J.; Marroyo, L. Modeling and control of n-paralleled grid-connected inverters with LCL filter coupled due to grid impedance in PV plants. *IEEE Trans. Power Electron.* **2011**, *26*, 770–785. [CrossRef]
5. Hosseinzadeh, M.; Salmasi, F.R. Fault-tolerant supervisory controller for a hybrid AC/DC micro-grid. *IEEE Trans. Smart Grid* **2018**, *9*, 2809–2823. [CrossRef]
6. Zhang, D.; Wang, F.; Burgos, R.; Boroyevich, D. Common-mode circulating current control of paralleled interleaved three-phase two-level voltage-source converters with discontinuous space-vector modulation. *IEEE Trans. Power Electron.* **2011**, *26*, 3925–3935. [CrossRef]
7. Prodanovic, M.; Green, T.C. High-quality power generation through distributed control of a power park microgrid. *IEEE Trans. Ind. Electron.* **2006**, *53*, 1471–1482. [CrossRef]

8. Hosseinzadeh, M.; Salmasi, F.R. Power management of an isolated hybrid AC/DC micro-grid with fuzzy control of battery banks. *IET Renew. Power Gener.* **2015**, *9*, 484–493. [CrossRef]

9. Hosseinzadeh, M.; Salmasi, F.R. Robust optimal power management system for a hybrid AC/DC micro-grid. *IEEE Trans. Sustain. Energy* **2015**, *6*, 675–687. [CrossRef]

10. Yu, C.; Zhang, X.; Liu, F.; Li, F.; Xu, H.; Cao, R.; Ni, H. Modeling and resonance analysis of multiparallel inverters system under asynchronous carriers conditions. *IEEE Trans. Power Electron.* **2017**, *32*, 3192–3205. [CrossRef]

11. He, J.; Li, Y.W.; Bosnjak, D.; Harris, B. Investigation and active damping of multiple resonances in a parallel-inverter-based microgrid. *IEEE Trans. Power Electron.* **2013**, *28*, 234–246. [CrossRef]

12. Enslin, J.H.R.; Heskes, P.J.M. Harmonic interaction between a large number of distributed power inverters and the distribution network. *IEEE Trans. Power Electron.* **2004**, *19*, 1586–1593. [CrossRef]

13. Turner, R.; Walton, S.; Duke, R. Stability and bandwidth implications of digitally controlled grid-connected parallel inverters. *IEEE Trans. Power Electron.* **2010**, *57*, 3685–3694. [CrossRef]

14. Zheng, C.; Zhou, L.; Xie, B.; Zhang, Q.; Li, H. A stabilizer for suppressing harmonic resonance in multi-parallel inverter system. In Proceedings of the 2017 IEEE Transportation Electrification Conference and Expo, Asia-Pacific, ITEC Asia-Pacific 2017, Harbin, China, 7–10 August 2017.

15. Wang, X.; Blaabjerg, F.; Liserre, M.; Chen, Z.; He, J.; Li, Y. An active damper for stabilizing power-electronics-based ac systems. *IEEE Trans. Power Electron.* **2014**, *29*, 3318–3329. [CrossRef]

16. Yuan, X.; Wang, F.; Boroyevich, D.; Li, Y.; Burgos, R. Dc-link voltage control of a full power converter for wind generator operating in weak-grid systems. *IEEE Trans. Power Electron.* **2009**, *24*, 2178–2192. [CrossRef]

17. Yuan, X.; Chai, J.; Li, Y. Control of variable pitch, variable speed wind turbine in weak grid systems. In Proceedings of the Energy Conversion Congress and Exposition, Atlanta, GA, USA, 12–16 September 2010; pp. 3778–3785.

18. Nawir, M.; Adeuyi, O.D.; Wu, G.; Liang, J. Voltage stability analysis and control of wind farms connected to weak grids. In Proceedings of the 13th IET International Conference on AC and DC Power Transmission (ACDC 2017), Manchester, UK, 14–16 February 2017; pp. 1–6.

19. Li, M.; Zhang, X.; Yang, Y.; Cao, P. The grid impedance adaptation dual mode control strategy in weak grid. Presented at the 2018 International Power Electronics Conference, Niigata, Japan, 20–24 May 2018.

20. Liserre, M.; Teodorescu, R.; Blaabjerg, F. Stability of photovoltaic and wind turbine grid-connected inverters for a large set of grid impedance values. *IEEE Trans. Power Electron.* **2006**, *21*, 263–272. [CrossRef]

21. Pan, D.; Ruan, X.; Bao, C.; Li, W.; Wang, X. Capacitor-current-feedback active damping with reduced computation delay for improving robustness of LCL-type grid-connected inverter. *IEEE Trans. Power Electron.* **2014**, *29*, 3414–3427. [CrossRef]

22. Guerrero, J.M.; Luis Garcia de, V.; Matas, J.; Castilla, M.; Miret, J. Output impedance design of parallel-connected ups inverters with wireless load-sharing control. *IEEE Trans. Ind. Electron.* **2005**, *52*, 1126–1135. [CrossRef]

23. Asiminoaei, L.; Teodorescu, R.; Blaabjerg, F.; Borup, U. A digital controlled PV-inverter with grid impedance estimation for ens detection. *IEEE Trans. Power Electron.* **2005**, *20*, 1480–1490. [CrossRef]

24. Cai, W.; Liu, B.; Duan, S.; Zou, C. An islanding detection method based on dual-frequency harmonic current injection under grid impedance unbalanced condition. *IEEE Trans. Ind. Inform.* **2013**, *9*, 1178–1187. [CrossRef]

25. Roinila, T.; Messo, T. Online grid-impedance measurement using ternary-sequence injection. *IEEE Trans. Ind. Appl.* **2018**, *1*. [CrossRef]

26. Hoffmann, N.; Fuchs, F.W. Minimal invasive equivalent grid impedance estimation in inductive-resistive power networks using extended kalman filter. *IEEE Trans. Power Electron.* **2014**, *29*, 631–641. [CrossRef]

27. Liserre, M.; Blaabjerg, F.; Teodorescu, R. Grid impedance estimation via excitation of LCL-filter resonance. *IEEE Trans. Ind. Appl.* **2007**, *43*, 1401–1407. [CrossRef]

energies

MDPI

Article

VSG-Based Dynamic Frequency Support Control for Autonomous PV–Diesel Microgrids

Rongliang Shi and Xing Zhang *

School of Electrical Engineering and Automation, Hefei University of Technology, Hefei 230009, China; shirl163@163.com
* Correspondence: honglf@ustc.edu.cn; Tel.: +86-551-6290-1408; Fax: +86-551-6290-1408

Received: 3 June 2018; Accepted: 5 July 2018; Published: 11 July 2018

Abstract: This paper demonstrates the use of a novel virtual synchronous generator (VSG) to provide dynamic frequency support in an autonomous photovoltaic (PV)–diesel hybrid microgrid with an energy storage system (ESS). Due to the lack of enough rotating machines, PV fluctuation might give rise to unacceptable frequency excursions in the microgrid. The VSG entails controlling the voltage-source inverter (VSI) to emulate a virtual inertial and a virtual damping via power injection from/to the ESS. The effect of the VSG on the frequency is investigated. The virtual inertia decreases the maximum frequency deviation (MFD) and the rate of change of frequency (RoCoF), but in exchange for raising the virtual inertia, the system is more oscillating. Meanwhile, raising the virtual damping brings reductions in the amplitude of the oscillations of frequency. However, the dynamic frequency support provided by them is lagging behind. In this regard, an improved VSG based on the differential feedforward of the diesel generator set (DGS) output current is proposed to further mitigate the MFD and the RoCoF. Simulations and experimental results from an autonomous microgrid consisted of a 400 kW DGS, and a 100 kVA VSG are provided to validate the discussion.

Keywords: virtual synchronous generator; dynamic frequency support; autonomous microgrid; virtual inertia; virtual damping; differential feedforward

---

1. Introduction

Photovoltaic (PV)–diesel hybrid microgrids are a fine choice for electricity generation in isolated regions where PV resource is rich [1]. The microgrid is constructed by a reduced number of controlled diesel generator sets (DGSs), and solar resources are used to supplement power generation [2]. However, the frequency excursions are more likely to happen in autonomous microgrids than in large interconnected power grids, since they feature rapid changes in the power demands and a relatively small generation capacity, especially in a high penetration power system with many intermittent renewable generators [3]. Furthermore, if DGS units cannot keep the maximum frequency deviation (MFD) as well as the rate of change of frequency (RoCoF) within stipulated operating ranges, skipping of power generators and loads may occur [4]. Hence, the help of energy storage systems (ESSs) is required to keep the dynamic stability of frequency for the self-existent microgrid.

Different solutions can be found in the control strategies of a voltage-source inverter (VSI) with an ESS. An ESS-based dynamic frequency support solution is used to cope with system frequency deviation by filtering of the output power of stochastic resources [5]. Nevertheless, this strategy needs the measurement of the output power, which requires a communication bus. The droop method is used to control the distributed energy storage interfaces independently by the local measurements, without a communication bus [6–10]. However, this strategy is aimed to provide only frequency adjustment by employing the permanent frequency droop form, thus, it cannot deal with the problem of dynamic frequency regulation. In virtual synchronous generator (VSG) method, which is VSI based, the ESS is managed in a way to emulate a virtual inertia and a virtual damping of a synchronous generator (SG) [11–18].

The constant parameters (CP)-VSG used to provide dynamic frequency support involves exhibiting the damping power and the inertial response of a real conventional SG [19,20]. The emulation of inertial response generally involves the domination of power that is inversely proportional to the first time derivative of the frequency [21]. On the other hand, the virtual damping control helps to weaken fluctuations, and then, decrease the settling time of the frequency [22]. Nevertheless, they do not investigate the use of the adjustable virtual parameters that can adjust their values during operation. As a result, the self-tuning (ST)-VSG with the variable virtual parameters is operating as a controlled current source, which is used to attenuate the frequency fluctuations [4]. However, the online optimization is asked to maintain the virtual damping power and the inertial response, which raises the calculation amount of the digital signal processor (DSP). Whereas, for the microgrid applications, especially considering the islanding mode, the VSG is selected to run with a voltage control method (VCM), as it is able to provide enough frequency and voltage support for the loads [23]. In this regard, the bang–bang control strategy with the variable virtual inertia is proposed to attenuate the frequency fluctuations [24]. The sign of the derivative of angular velocity with respect to the sign of the relative angular velocity defines the acceleration or deceleration. They act in the same direction, therefore, it is an acceleration period. In addition, as they have opposite signs, it is a deceleration. By choosing a large value for the moment of inertia during acceleration, the haste is decreased. During deceleration, a small value for inertia factor is used to improve the deceleration effect. In other words, the virtual inertia is changed between a large value of inertia and a small one for four times during each cycle of fluctuations. Unfortunately, using a large constant virtual inertia for the bang–bang control will lead to a sluggish response. In addition, each switching may cause the power fluctuations.

Despite the effectiveness of the above techniques, the inertial response, as well as the damping power of the CP-VSG or the ST-VSG, are realized by measuring both the frequency variations and the frequency deviations from the nominal grid frequency value. As a result, the dynamic frequency support provided by the inertial response as well as the damping power is behind the variations of loads. Thus, the effect of improving the frequency stability using the inertial response and the damping power is limited. In this regard, an improved VSG based on the differential feedforward of the DGS output current is proposed to further mitigate the MFD and the RoCoF.

This paper is organized as follows. In Section 2, the structure of an autonomous microgrid and the control model of a DGS are reviewed. Section 3 analyzes the impacts of the inertial response as well as the damping power of the VSG on the frequency stability. Section 4 explains the proposed VSG with the differential feedforward of the DGS output current and its performance evaluation. In Section 5, simulations and experimental results from an autonomous microgrid consisting of a 400 kW DGS and a 100 kVA VSG are demonstrated. Section 6 is devoted to the conclusion.

2. System Overview

Figure 1 shows a typical system structure of an autonomous PV–battery–diesel microgrid based on a PV generator, two VSI units, and a DGS working in parallel. For each space vector pulse width modulation (SVPWM) VSI unit, the dc power is supplied by an ESS, and a fixed dc voltage is assumed in this paper to simplify the theoretic analysis. Each VSI unit is integrated to the ac-bus through a LC filter and a distribution transformer (DT). The VSI can inject power either into the ac-bus—inverter mode, or into the ESS—rectifier mode. Depending on the proportion between PV and ESS power, the microgrid operation can be characterized as a low PV penetration level and a high PV penetration level.

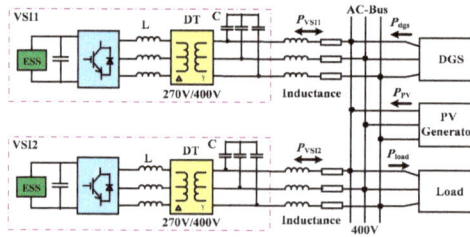

Figure 1. Typical structure of an autonomous photovoltaic (PV)–battery–diesel microgrid.

Based on a DGS unit and two VSI units running in parallel, the power contribution of the PV generator is relatively small, i.e., the autonomous microgrid operates with a low PV penetration. In this situation, frequency variations due to PV fluctuations would be in an appropriate limit. The active power balance in the ac-bus is given by

$$P_{VSI1} + P_{VSI2} + P_{dgs} + P_{PV} = P_{load},$$ (1)

where P_{VSI1}, P_{VSI2}, P_{dgs}, P_{PV}, and P_{load} are the VSI units, DGS, PV generator, and load instantaneous active powers, respectively.

On the other hand, for a certain load condition, only one VSI and a DGS operating in isochronous mode will be service. Under this circumstance, the power dedication of the solar source might be comparatively high, i.e., the microgrid would run in high PV penetration. As a result, PV fluctuations would produce unacceptable frequency variations. However, the frequency variations result in rapid acceleration or deceleration of the DGS rotor, and they can be mitigated with a suitable control method. The active power balance in the ac-bus is given by

$$P_{VSI} + P_{dgs} + P_{PV} = P_{load},$$ (2)

where P_{VSI} is the active power of the VSI in service.

As illustrated before, this paper concentrates on the situation when the autonomous microgrid is exposed to a high PV penetration. For purpose of analysis, it is assumed that the voltage amplitude of the ac-bus is regulated, and it is considered constant.

The DGS model is made up of two major components coupled by a collaborative shaft: the diesel prime mover and the SG. Due to the scope of this paper, the speed control schematic diagram of the DGS with a proportional–integrative (PI) controller is shown in Figure 2, according to [1].

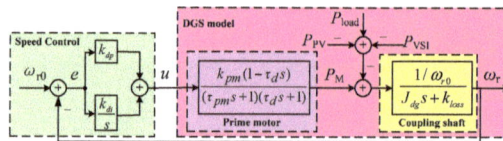

Figure 2. Diesel generator set (DGS) speed control based on a proportional–integrative (PI) controller.

In Figure 2, J_{dg} is the rotating inertia, ω_r is the rotor angular speed, and k_{loss} represents the rotational losses, ω_{r0} is the synchronous speed, P_M is the mechanical power supplied by the prime motor, k_{pm} concentrates the gain of fuel injection system and diesel engine, τ_{pm} is the time constant of the fuel injection system, τ_d is the dead time of the diesel engine, k_{dp} and k_{di} are the proportional and integral factors of a PI controller. The parameters of a 26 kW DGS obtained from [25] are shown in Table 1.

Table 1. The main parameters of the diesel generator set (DGS).

Parameter	Value	Parameter	Value
k_{pm}	1	τ_d	0.011 s
ω_{r0}	314.16 rad/s	k_{loss}	0.02 kg·m^2/s
J_{dg}	0.66 kg·m^2	τ_{pm}	0.2 s
k_{dp}	409.5	k_{di}	367.3

3. Impact of Virtual Synchronous Generator (VSG) on Frequency Stability

This section presents the utilization of a VSG to provide dynamic frequency control in an autonomous microgrid. Particularly, the VSG is realized with a VSI, and the method used to provide dynamic frequency support consists of simulating the damping power and the inertial response of a real SG, which are available only during a transient.

If the active power of the VSG is managed in inverse proportion to the derivative of the rotor speed, then promoting the inertial response of the DGS to variations in the power demand [4]. Hence, the active power of the VSG is defined as

$$P_{VSI} = -J\omega_{r0}\frac{d\omega_r}{dt},\tag{3}$$

where J is the virtual inertia. The inertial response of the VSG simulates the power that is inherently absorbed or released by a real SG as the power demand changes [21]. When the rotor speed begins to reduce (a negative derivative), the VSG begins to inject active power into the system until the rotor speed attains its minimum (when the first derivative is zero), then the rotor speed begins to rise (a positive derivative), and then the VSG begins to absorb active power. This procedure will last until the new steady state is realized. Figure 3 shows the block diagram for speed control with the virtual inertia control loop.

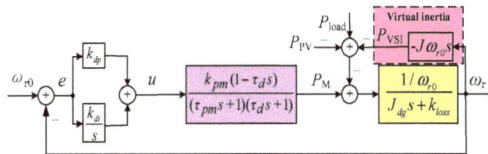

Figure 3. DGS speed control with the virtual inertia control loop.

From Figure 3, the transfer function in terms of the active powers as inputs can be obtained as follows:

$$G_J(s) = \frac{\omega_r(s)}{P_{PV}(s) - P_{load}(s)} = \frac{s}{(J_{eq}\omega_{r0} - k_{dp}\tau_d)s^2 + Ms + k_{di}},\tag{4}$$

where $J_{eq} = (J_{dg} + J)$ is the total inertia, $M = (k_{loss}\omega_{r0} + k_{dp} - k_{di}\tau_d)$. On the basis of (4), the standard parameters of a second-order transient response with the virtual inertia loop can be defined as

$$\xi_1 = \frac{M}{2\sqrt{k_{di}(J_{eq}\omega_{r0} - \tau_d k_{dp})}}, \quad \omega_{n1} = \frac{k_{di}}{\sqrt{k_{di}(J_{eq}\omega_{r0} - \tau_d k_{dp})}},\tag{5}$$

where ω_{n1} is the natural oscillation frequency, ξ_1 is the damping ratio. From the (5), it could be established that, in exchange for adding the virtual inertia, the control system becomes more oscillatory and slower as the values of ξ_1 and ω_{n1} become smaller. Due to space limitations, the design of the virtual inertia is not discussed here, and more details are available in [26].

On the other hand, if the active power of the VSG is managed in proportion to the deviation of the rotor speed, then enhancing the damping power of the DGS leads to variations in the power demand. The active power of the VSG is defined as

$$P_{VSI} = D\omega_0(\omega_{r0} - \omega_r), \tag{6}$$

where D is the virtual damping of the VSG. The damping power is supplied by the virtual damper windings that help to weaken oscillations, and hence, reduce the stabilization time for a predefined tolerance band. Any deviation from the synchronous speed generates a power that tries to bring back the rotor speed to the reference [22]. Figure 4 shows the block diagram for speed control with the virtual damping control loop.

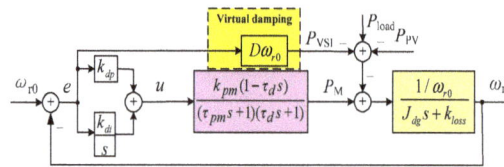

Figure 4. DGS speed control with the virtual damping control loop.

From Figure 4, the transfer function in terms of the active powers as inputs can be obtained as follows:

$$G_D(s) = \frac{\omega_r(s)}{P_{PV}(s) - P_{load}(s)} = \frac{s}{(J_{dg}\omega_{r0} - k_{dp}\tau_d)s^2 + Ns + k_{di}}, \tag{7}$$

where $N = [(k_{loss} + D)\omega_{r0} + k_{dp} - k_{di}\tau_d]$. On the basis of (7), the standard parameters of a second-order transient response with the virtual damping loop can be defined as

$$\xi_2 = \frac{N}{2\sqrt{k_{di}(J_{dg}\omega_{r0} - \tau_d k_{dp})}}, \quad \omega_{n2} = \frac{k_{di}}{\sqrt{k_{di}(J_{dg}\omega_{r0} - \tau_d k_{dp})}}, \tag{8}$$

where ω_{n2} is the natural oscillation frequency, ξ_2 is the damping ratio. From (8), it could be concluded that more virtual damping would assist to stabilize the control system faster as the value of ξ_2 becomes bigger.

In order to evaluate the above analysis results, the control block diagrams of the simulation of an autonomous microgrid compensated with the VSG were built according to Figures 3 and 4. The tests consist of a step increase in the power output of the PV generator, P_{PV}, at $t = 8$ s from 0 kW to 10 kW, while the load power demand, P_{load}, remains constant at 15 kW. It should be noted that the PV production is connected to the microgrid by a PV inverter which is operating as a controlled current source. Then, a step increase in the power output of the PV generator can be achieved by regulating the PV inverter in this paper. In these conditions, the microgrid will operate at a high PV penetration level of 40%. The DGS parameters used in simulations are presented in Table 1. The suitable values of J and D can be determined by the (5) and (8) using the parameters listed in Table 1. Simulations are made for speed control without virtual inertia, $J = 0$, and with virtual inertia control, $J = 0.32$ kg·m^2 and $J = 0.64$ kg·m^2.

From Figure 5a, it can be seen that the key effect of increasing virtual inertia is that the MFD decreases. However, a side effect of increasing virtual inertia is that the frequency will oscillate for a longer time. Meanwhile, increasing the virtual damping obtains a reduction in the MFD, as can be seen from Figure 5b.

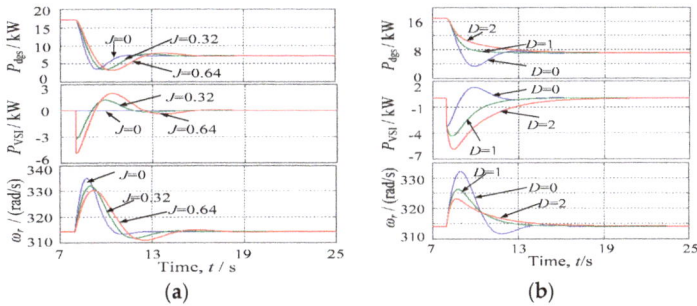

Figure 5. Effects of the inertia and damping on an active power transient. (a) Different values of virtual inertia and $D = 0$; (b) Different values of virtual damping and $J = 0.32$ kg·m^2.

As mentioned before, the virtual inertia as well as the virtual damping are performed by means of the ESS. Figure 5 also shows that the values of J and D are directly related to the expected dynamic performance and proper capacity of the ESS.

The angular frequency-acceleration trajectories of simulated DGS are shown in Figure 6. The advantage of adding the virtual inertia is the reduction of the RoCoF due to a power disturbance. For $J = 0.64$ kg·m^2, the maximum RoCoF is reduced from 8 Hz/s to 4 Hz/s, as can be seen from Figure 6a. From Figure 6b, it can be seen that the advantages of increasing the virtual damping are the attenuation of the amplitude of the oscillations of the frequency and the reduction of the MFD. For $D = 2$, the MFD is reduced from 2.85 Hz to 1.4 Hz.

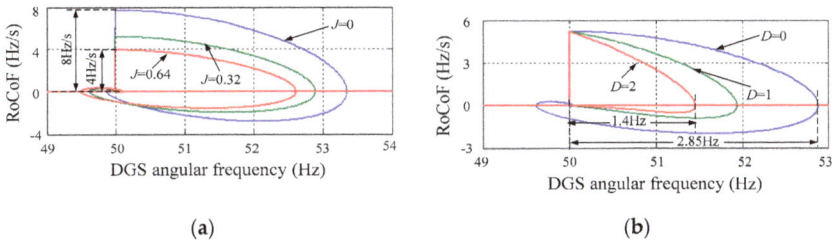

Figure 6. Angular frequency-acceleration curves of DGS with (a) virtual inertia control loop; (b) virtual damping control loop.

It is important to point out that, the variations of loads mainly result in the oscillations of the frequency. As a result, the changes of the latter lag behind the changes of the former. Since the inertial response as well as the damping power of the VSG are realized by measuring the variations of the frequency, thus, the dynamic frequency support provided by them is also lagging behind. In this regard, alternative approaches to provide a dynamic power response that can be synchronized with the variations of loads or even be ahead of the variations of loads.

4. Proposed VSG and Its Performance Evaluation

Considering that the first frequency oscillation is the most serious one in terms of maintaining the system frequency stability, an improved VSG based on the differential feedforward of the DGS output current is proposed to further decrease the MFD and the RoCoF. Note that the DGS output power can be approximately equal to its mechanical power under the condition of neglecting the mechanical loss and the iron loss. As a result, the differential of the DGS output current can be replaced by the differential of its mechanical power, assuming that the voltage amplitude on the ac-bus is fixed.

In this case, the active power of the VSG based on the differential feedforward of the DGS output current is defined as

$$P_{VSI} = \frac{k_{df}s}{\tau s + 1}P_{dgs} \approx \frac{k_{df}s}{\tau s + 1}P_M, \tag{9}$$

where k_{df} is the differential feedforward gain, and τ is the filtering time gain of the low-pass filter (LPF). The active power is typically calculated from the time derivative of the DGS mechanical power that can expand the noise that is usually contained in this power. Hence, a LPF is applied to prevent the excessive noise from polluting the system. Figure 7 is the block diagram for speed control with the differential feedforward of the DGS output current control loop.

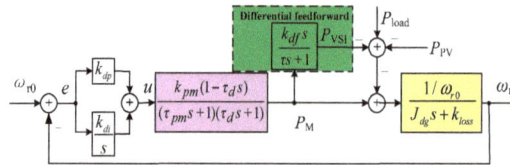

Figure 7. DGS speed control with the differential feedforward of its output current control loop.

From Figure 7, the transfer function in terms of the active powers as inputs can be obtained as follows:

$$G_{DF}(s) = \frac{\omega_r(s)}{P_{PV}(s) - P_{load}(s)} = \frac{(\tau s + 1)s}{as^3 + bs^2 + cs + k_{di}}, \tag{10}$$

where

$$\begin{cases} a = J_{dg}\omega_{r0}\tau - (\tau + k_{df})k_{dp}\tau_d \\ b = (k_{loss}\tau + J_{dg})\omega_{r0} - k_{dp}\tau_d + (k_{dp} - k_{di}\tau_d)(\tau + k_{df}) \\ c = k_{loss}\omega_{r0} + k_{dp} - k_{di}\tau_d + (\tau + k_{df})k_{di} \end{cases} \tag{11}$$

Using (10) and (11), the trajectory of the three eigenvalues ($\lambda_1, \lambda_2, \lambda_3$) can be obtained. Figure 8a shows the trajectory of these eigenvalues as a function of the differential feedforward gain k_{df}. It can be seen that as k_{df} is increased, conjugate eigenvalues λ_1 and λ_2 move towards the real axis, making the system less oscillatory. Eigenvalue λ_3 moves towards the stable region, which makes the system evolve from a third-order system to an approximate second-order system. Meanwhile, Figure 8b shows the trajectory of these eigenvalues as a function of the filter time gain τ. It can be seen that as τ is increased, conjugate eigenvalues λ_1 and λ_2 move towards the real axis, making the system less oscillatory. Eigenvalue λ_3 moves towards the unstable region, which makes the system evolve from a second-order system to an approximate third-order system. Figure 8a,b have been obtained with the DGS parameters given in Table 1.

It is to be noted that a LPF with high cut-off frequency is required to enhance the dynamic response of the VSG, whereas a large filter time gain is required to obtain good attenuation of high frequency distortion components in the calculated active power and to increase the system damping. For this reason, and to operate without the distortion components, it is necessary to use larger τ values, like $\tau = 0.3$ s, or $\tau = 0.5$ s, but in exchange for longer dynamic reactions.

Based on the above analysis, simulations are made for speed control without differential feedforward, $k_{df} = 0$, and with differential feedforward control, $k_{df} = 1$ and $k_{df} = 2$. From Figure 9a, it can be seen that adding the differential feedforward gain produces a reduction in the MFD. Adding differential feedforward gain mitigates the MFD and the RoCoF effectively, due to a power disturbance. For $k_{df} = 2$, the MFD is reduced from 2.85 Hz to 0.85 Hz, and the RoCoF is decreased quickly, as can be seen from Figure 9b.

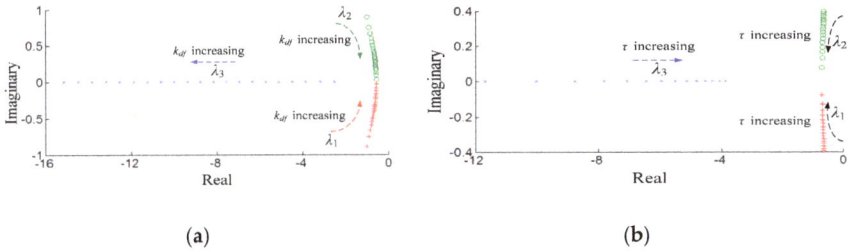

Figure 8. Traces of the three eigenvalues as a function of (**a**) differential feedforward gain: $\tau = 0.3$, $0 \le k_{df} \le 2$; (**b**) filter time gain: $k_{df} = 1$, $0.1 \le \tau \le 0.96$.

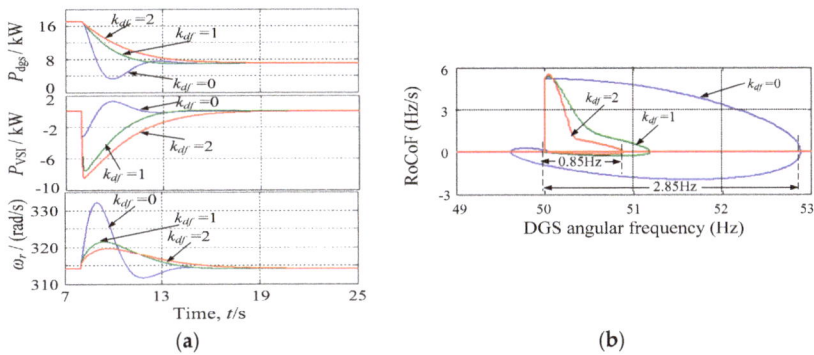

Figure 9. (**a**) Effects of the different differential feedforward gains ($D = 0$. and $J = 0.32$ kg·m²) on an active power transient; (**b**) Angular frequency-acceleration curves of DGS with differential feedforward control loop.

The proposed control method for the DGS is implemented as shown in Figure 10. This control scheme can provide virtual inertia, virtual damping, and differential feedforward control for the DGS to support dynamic frequency control.

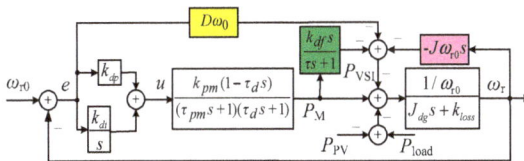

Figure 10. Proposed control strategy for the DGS to support dynamic frequency control.

With Figure 10, the angular frequency-acceleration trajectories of simulated DGS are shown in Figure 11. It is obvious that the MFD and the RoCoF are mitigated more effectively for $J = 0.32$ kg·m², $D = 1$, $k_{df} = 2$ than for $J = 0.32$ kg·m², $D = 1$, $k_{df} = 1$, or for $J = 0.32$ kg·m², $D = 1$, $k_{df} = 0$, or for $J = 0.32$ kg·m², $D = 0$, $k_{df} = 0$. It implies that the proposed VSG with the differential feedforward control loop is more effective in enhancing the system frequency stability than the VSG without the differential feedforward control loop.

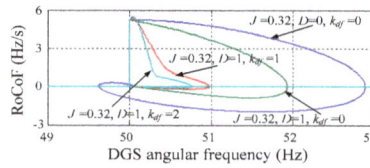

Figure 11. Angular frequency-acceleration curves of DGS with virtual inertia, virtual damping, or differential feedforward control loop.

5. Simulation and Experimental Results

The proposed VSG control strategy for the DGS to support dynamic frequency control is confirmed through Matlab/Simulink (R2011b, MathWorks, Natick, MA. USA), and experimentally on an autonomous microgrid. Figure 12 describes the schematic of the simulation and experiment of the microgrid consisted of a 100 kVA VSG and a 400 kW DGS for a load step of 100 kW at $t = 3$ s. The system parameters of the simulation, as well as the experiments, are almost the same, and are enumerated in Table 2.

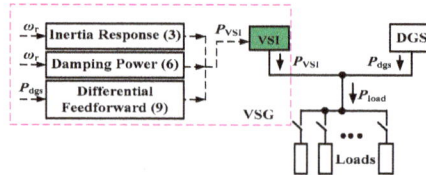

Figure 12. An autonomous microgrid with a DGS and a virtual synchronous generator (VSG) unit.

Table 2. Parameters in the simulation and experiment.

The 100 kVA VSG System		
Parameter	Simulation	Experiment
System voltage	380 V, 50 Hz	380 V, 50 Hz
DC link voltage	600 V	600 V
LC filter	L = 0.56 mH, C = 270 μF	L = 0.56 mH, C = 270 μF
Power reference	P_{ref} = 20 kW, Q_{ref} = 0	P_{ref} = 20 kW, Q_{ref} = 0
Switching frequency	5 kHz	5 kHz
The 400 kW DGS System		
Parameter	Simulation	Experiment
System voltage	380 V, 50 Hz	380 V, 50 Hz
Total inertia	3.6 kg·m^2	3.6 kg·m^2
Rotational loss	0.24 kg·m^2/s	0.24 kg·m^2/s
Fuel injection time	60 ms	60 ms
Engine delay	15 ms	15 ms
Governor gains	K_p = 0.16, K_i = 0.9	K_p = 0.16, K_i = 0.9
AVR gains	k_p = 0.12, k_i = 0.12	k_p = 0.12, k_i = 0.12

A. Simulation verification

(1) Virtual inertia: From Figure 13a, it can be seen that a side effect of increasing virtual inertia is that the system frequency will oscillate for a longer time. However, increasing virtual inertia reduces the MFD and the RoCoF effectively due to a power disturbance. For $J = 6$ kg·m^2, the MFD is decreased from 0.5 Hz to 0.45 Hz and the maximum RoCoF is decreased from 3.2 Hz/s to 2 Hz/s, as can be seen from Figure 13b.

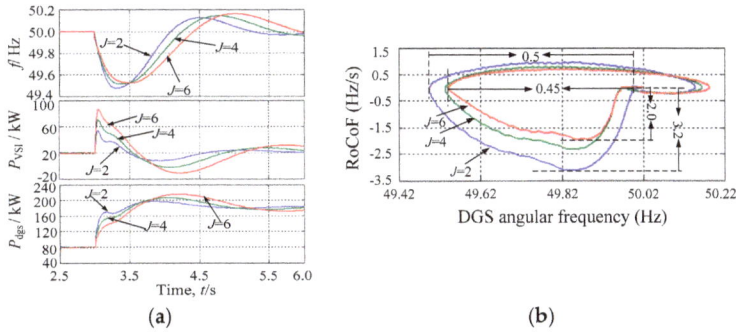

Figure 13. (**a**) Effects of the different values of virtual inertia and zero virtual damping on an active power transient; (**b**) Angular frequency-acceleration curves of DGS with virtual inertia control loop.

(2) Virtual damping: The same tests are conducted for the VSG with the added virtual damping to support dynamic frequency control on an active power transient. Figure 14a displays the simulation results for the alterations of the system frequency, the output active powers of the VSG, and the DGS. It is observed that the advantages of adding virtual damping to the system are the attenuation of the amplitude of the oscillations of the frequency and the reduction of the MFD.

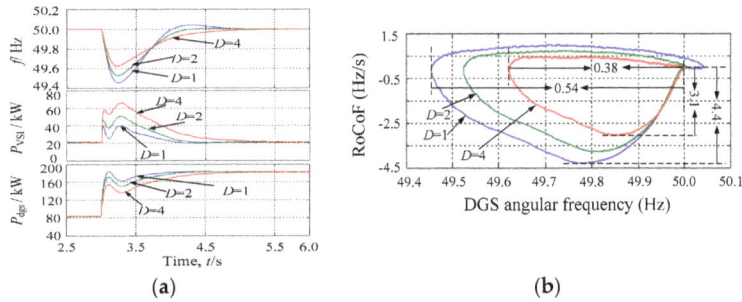

Figure 14. (**a**) Effects of the different values of virtual damping and zero virtual inertia on an active power transient; (**b**) Angular frequency-acceleration curves of DGS with virtual damping control loop.

From Figure 14b, it can be seen that the curve "$D = 4$" illustrates the RoCoF and the MFD with respect to the rated value that situates inside the other two curves. It explains that increasing the virtual damping produces reductions in the MFD and the RoCoF. For $D = 4$, the MFD is reduced from 0.54 Hz to 0.38 Hz, and the maximum RoCoF is reduced from 4.4 Hz/s to 3.1 Hz/s.

(3) Differential feedforward: Figure 15a displays the simulation results for the alterations of the system frequency, the output active powers of the VSG, and the DGS. It is observed that the advantages of adding the differential feedforward gain are the attenuation of the amplitude of the oscillations of the frequency and the reduction of the MFD. On the other hand, the curve "$k_{df} = 4$" illustrates the RoCoF and the MFD with respect to the rated value that situates inside the other two curves. It explains that increasing the differential feedforward gain mitigates both the MFD and the RoCoF effectively. For $k_{df} = 4$, the MFD is reduced from 0.45 Hz to 0.29 Hz, and the maximum RoCoF is reduced from 2.3 Hz/s to 1.1 Hz/s, as can be seen from Figure 15b.

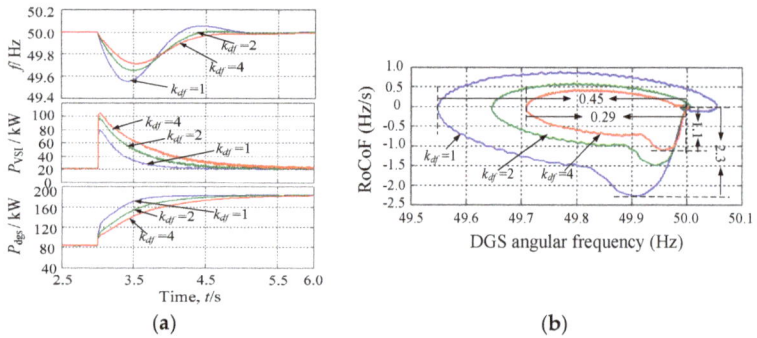

Figure 15. (**a**) Effects of the different differential feedforward gains ($D = 0$ and $J = 0$) on an active power transient; (**b**) Angular frequency-acceleration curves of DGS with differential feedforward control loop.

B. Experimental verification

Experiments are also carried out on an autonomous microgrid consisting of a 100 kVA VSG and a 400 kW DGS for a load step of 100 kW to verify the proposed VSG-based dynamic frequency support scheme. In the experiments, a three-phase programmable rectifier is used to represent the ESS, and the three-phase VSG system is controlled by a TMS320F28335 DSP (TMS320C28x, Texas Instruments, Dallas, TX, USA), which implements the proposed control strategies, as described in the previous sections. The VSG unit consists of a three-phase IGBT full bridge with a switching frequency of 5 kHz and an LC output filter, using the parameters listed in Table 2. The experimental platform of the autonomous microgrid is shown in Figure 16. The transitory regime is created by switching a resistive load of 100 kW on at certain time intervals in the experiment as shown in Figure 16a.

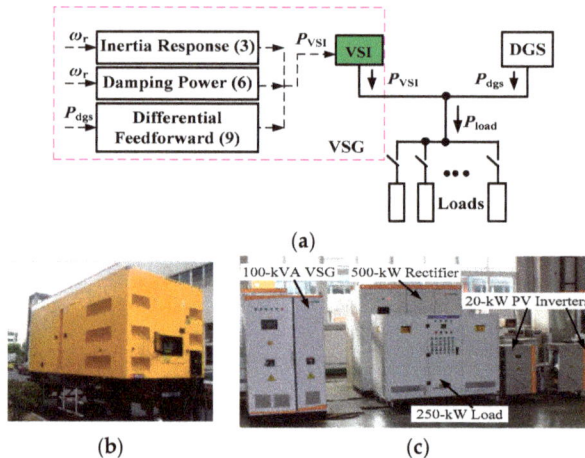

Figure 16. Experimental platform. (**a**) An autonomous microgrid with a DGS and a VSG unit; (**b**) A 400 kW DGS; (**c**) The VSG system.

(1) Virtual inertia: The performance of increasing virtual inertia is shown in Figure 17a, where it can be found that the system frequency oscillates for a longer time. When the dynamic frequency support is complemented by the virtual inertia control loop (with $J = 2$ kg·m^2, $J = 4$ kg·m^2, and

$J = 6$ kg·m^2), the VSG outputs dynamic active power. As a result, both the MFD and the RoCoF are decreased, as shown in Figure 17b. For $J = 6$ kg·m^2, the MFD is decreased from 0.58 Hz to 0.53 Hz, and the maximum RoCoF is decreased from 8.4 Hz/s to 6.0 Hz/s.

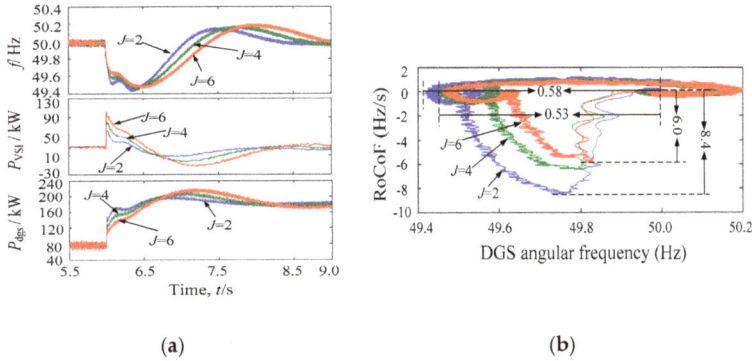

(a) (b)

Figure 17. (a) Experimental results of the different values of virtual inertia and zero virtual damping on an active power transient; (b) Angular frequency-acceleration curves of DGS with virtual inertia control loop.

(2) Virtual damping: Figure 18a displays the experimental results for the alterations of the system frequency, the output active powers of the VSG, and the DGS. It is observed that the advantages of increasing virtual damping are the attenuation of the amplitude of the oscillations of the frequency and the reduction of the MFD.

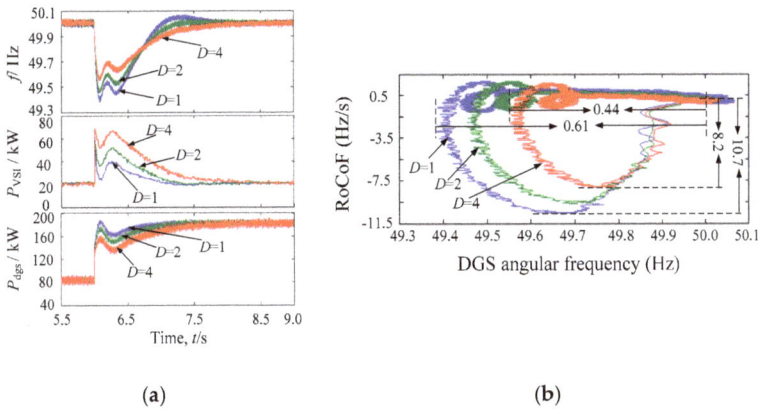

(a) (b)

Figure 18. (a) Experimental results of the different values of virtual damping and zero virtual inertia on a power transient; (b) Angular frequency-acceleration curves of DGS with virtual damping control loop.

When the dynamic frequency support is complemented by the virtual damping control loop (with $D = 1$, $D = 2$, and $D = 4$), the VSG outputs dynamic active power. As a result, both the MFD and the RoCoF are decreased, as shown in Figure 18b. For $D = 4$, the MFD is decreased from 0.61 Hz to 0.44 Hz, and the maximum RoCoF is decreased from 10.7 Hz/s to 8.2 Hz/s.

(3) Differential feedforward: Figure 19a displays the experimental results for the alterations of the system frequency, the output active powers of the VSG, and the DGS. It is observed that the advantages

of adding the differential feedforward gain are the attenuation of the amplitude of the oscillations of the frequency and the reduction of the MFD.

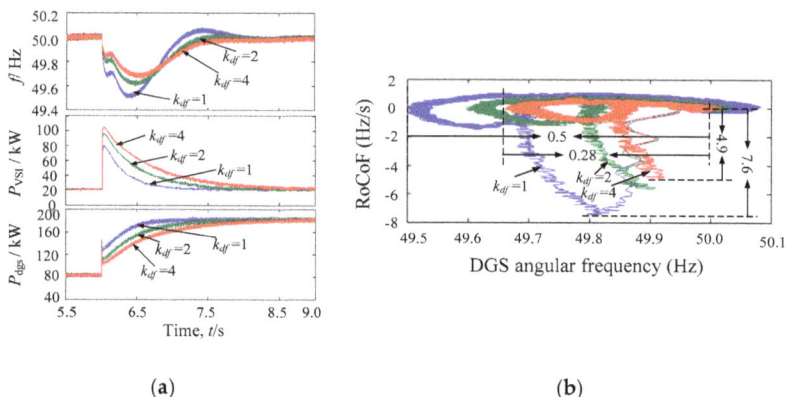

(a)   (b)

Figure 19. (a) Experimental results of the different differential feedforward gains ($D = 0$ and $J = 0$) on an active power transient; (b) Angular frequency-acceleration curves of DGS with differential feedforward control loop.

On the other hand, when the dynamic frequency support is complemented by the differential feedforward control loop (with $k_{df} = 1$, $k_{df} = 2$, and $k_{df} = 4$), the VSG outputs dynamic active power. As a result, both the MFD and the RoCoF are decreased, as shown in Figure 19b. For $k_{df} = 4$, the MFD is reduced from 0.5 Hz to 0.28 Hz, and the maximum RoCoF is reduced from 7.6 Hz/s to 4.9 Hz/s.

C. Summary

Table 3 summarizes the experimental results. Absolute error is calculated only between the maximum value and the minimum value as $(x_{max} - x_{min})$. To compare the performances of the virtual inertia, the virtual damping, and the differential feedforward, two indexes that give a survey of how much energy is applied from the ESS to decrease the MFD and the RoCoF are defined as

$$\begin{cases} I_{perf} = \dfrac{E_{AE}}{MFD_{AE}} = \dfrac{E_{de} - E_{re}}{MFD_{max} - MFD_{min}} [kJ/Hz] \\ I_{perdf} = \dfrac{E_{AE}}{RoCoF_{AE}} = \dfrac{E_{de} - E_{re}}{RoCoF_{max} - RoCoF_{min}} [kJ/(Hz/s)] \end{cases} \tag{12}$$

where E_{AE} is the absolute error between the delivered energy E_{de} and the recovered energy E_{re} of the VSG, MFD_{AE} is the absolute error of the MFD, and $RoCoF_{AE}$ is the absolute error of the RoCoF.

Table 3. Performance comparison among virtual inertia, virtual damping, and differential feedforward under load step.

Control Type	MFD_{AE} (Hz)	$RoCoF_{AE}$ (Hz/s)	E_{AE} (kJ)
Virtual inertia	0.05	2.4	11.3
Virtual damping	0.17	2.5	29.6
Differential feedforward	0.22	2.7	26.8

Table 4 displays the calculated results of I_{perf} and I_{perdf} for the VSG for the different experimental scenarios. It can be found that the virtual inertia control is more efficient in reducing the RoCoF than the other control methods, but it uses more energy in reducing the MFD. On the other hand, it can be found that the values calculated for the VSG with the differential feedforward control are continuously smaller than the values calculated for the VSG with the virtual damping control. This illustrates that the differential feedforward control uses less energy in reducing the MFD than the virtual damping control, using approximately 70% of the delivered energy. In addition, the differential feedforward control uses less energy in reducing the RoCoF than the virtual damping control, using approximately 84% of the delivered energy, i.e., is more effective in supporting dynamic frequency control.

Table 4. Performance indexes for virtual inertia, virtual damping, and differential feedforward under load step.

Control Type	I_{perf} (kJ/Hz)	I_{perdf} (kJ/(Hz/s))
Virtual inertia	226	4.71
Virtual damping	174.1	11.84
Differential feedforward	121.8	9.93

On the other hand, with the differential feedforward control, the system presented is more effective in reducing the MFD than the virtual inertia control, using approximately 54% of the delivered energy, however, it uses only 45% of energy in reducing the RoCoF used by the virtual inertia control.

6. Conclusions

The dynamic frequency support control for the autonomous PV–diesel hybrid microgrids with the ESS based on VSG is investigated in this paper. The performances of the virtual inertia, the virtual damping, as well as the differential feedforward are evaluated by comparing their reductions in the MFD and the RoCoF for the simulation and experimental scenarios. For experimental cases, the virtual inertia control is more efficient in reducing the RoCoF than the other control methods, but uses more energy in reducing the MFD. Meanwhile, a side effect of increasing virtual inertia is that the system becomes more oscillatory and slower. On the other hand, the differential feedforward control achieves the same performance to that of the virtual damping control while reducing the MFD and the RoCoF. Moreover, in all the experimental scenarios, the differential feedforward control uses less energy in reducing the MFD and the RoCoF than the virtual damping control, i.e., is more effective in supporting dynamic frequency control.

Finally, it was also found that, depending on the type of load variation, the operation of the proposed differential feedforward control may lead to a greater charge/discharge of the ESS when compared to the virtual inertia control. This suggests that further work is required in order to include the state of health and the state of charge of the ESS for the differential feedforward control. Also, a stability analysis is required in order to know the admissible values for k_{df}.

Author Contributions: R.S. designed the algorithm and performed the experiments and conduced simulations as the first author. X.Z. handled the project and contributed materials tools. All the authors approved the publication.

Funding: This research received no external funding.

Acknowledgments: This work was supported by the National key Research and Development Program of China (No. 2016YFB0900300), and the National Natural Science Foundation of China (No. 51677049).

Conflicts of Interest: The authors declare no conflict of interest.

Nomenclatures

P_{VSIi}	active power of the VSI unit (i = 1–2)
P_{dgs}	active power of the DGS
P_{PV}	active power of the PV generator
P_{load}	active power of the load
J_{dg}	rotating inertia of the DGS
ω_r	rotor angular speed
k_{loss}	rotational losses

ω_{r0}	synchronous speed
P_M	mechanical power
k_{pm}	gain of fuel injection system and diesel engine
τ_{pm}	time constant of the fuel injection system
τ_d	dead time of the diesel engine
k_{dp}	proportional factors of a PI controller
k_{di}	integral factors of a PI controller
J	virtual inertia
J_{eq}	total inertia
ω_{ni}	natural oscillation frequency ($i = 1$–3)
ξ_i	damping ratio ($i = 1$–3)
D	virtual damping
k_{df}	differential feedforward gain
τ	filtering time gain of the low-pass filter (LPF)
λ_i	conjugate eigenvalues ($i = 1$–3)
E_{AE}	absolute error
E_{de}	delivered energy
E_{re}	recovered energy
MFD_{AE}	absolute error of the MFD
$RoCoF_{AE}$	absolute error of the RoCoF
I_{perf}	energy index of the MFD
I_{perdf}	energy index of the RoCoF

References

1. Torres, M.; Lopes, L.A.C. Virtual synchronous generator control in autonomous wind-diesel power systems. In Proceedings of the IEEE Electrical Power & Energy Conference (EPEC), Montreal, QC, Canada, 22–23 October 2009.
2. Nacfaire, H. *Wind-Diesel and Wind Autonomous Energy Systems*; Elsevier Science: New York, NY, USA, 1989.
3. Committee E M. *IEEE Recommended Practice for Monitoring Electric Power Quality (Revision of IEEE Std 1159–1995)*; IEEE: New York, NY, USA, 2009.
4. Torres, L.M.A.; Lopes, L.A.; Moran, T.L.A.; Espinoza, C.J.R. Self-tuning virtual synchronous machine: A control strategy for energy storage systems to support dynamic frequency control. *IEEE Trans. Energy Convers.* **2014**, *29*, 833–840. [CrossRef]
5. Li, W.; Joos, G.; Abbey, C. Wind power impact on system frequency deviation and an ESS based power filtering algorithm solution. In Proceedings of the IEEE PES Power Systems Conference and Exposition, Atlanta, GA, USA, 29 October–1 November 2006.
6. Tao, Y.; Liu, Q.; Deng, Y.; Liu, X.; He, X. Analysis and mitigation of inverter output impedance impacts for distributed energy resource interface. *IEEE Trans. Power Electron.* **2015**, *30*, 3563–3576. [CrossRef]
7. Vasquez, J.C.; Guerrero, J.M.; Savaghebi, M.; Eloy-Garcia, J.; Teodorescu, R. Modeling, analysis, and design of stationary-reference-frame droop-controlled parallel three-phase voltage source inverters. *IEEE Trans. Ind. Electron.* **2013**, *60*, 1271–1280. [CrossRef]
8. Han, Y.; Shen, P.; Zhao, X.; Guerrero, J.M. An enhanced power sharing scheme for voltage unbalance and harmonics compensation in an islanded AC microgrid. *IEEE Trans. Energy Convers.* **2016**, *31*, 1037–1050. [CrossRef]
9. Guerrero, J.M.; Vasquez, J.C.; Matas, J.; Castilla, M.; Vicuña, L.G.D.; Castilla, M. Hierarchical control of droop-controlled AC and DC microgrids-A general approach toward standardization. *IEEE Trans. Ind. Electron.* **2011**, *58*, 158–172. [CrossRef]
10. Matas, J.; Castilla, M.; de Vicuna, L.G.; Miret, J.; Vasquez, J.C. Virtual impedance loop for droop-controlled single-phase parallel inverters using a second-order general-integrator scheme. *IEEE Trans. Power Electron.* **2010**, *25*, 2993–3002. [CrossRef]
11. Liu, J.; Miura, Y.; Ise, T. Comparison of dynamic characteristics between virtual synchronous generator and droop control in inverter- based distributed denerators. *IEEE Trans. Power Electron.* **2016**, *31*, 3600–3611. [CrossRef]

12. Shintai, T.; Miura, Y.; Ise, T. Oscillation damping of a distributed generator using a virtual synchronous generator. *IEEE Trans. Power Deliv.* **2014**, *29*, 668–676. [CrossRef]

13. Zhong, Q.-C.; Weiss, G. Synchronverters: Inverters that mimic synchronous generators. *IEEE Trans. Ind. Electron.* **2011**, *58*, 1259–1267. [CrossRef]

14. Bevrani, H.; Ise, T.; Miura, Y. Virtual synchronous generators: A survey and new perspectives. *Int. J. Electr. Power Energy Syst.* **2014**, *54*, 244–254. [CrossRef]

15. Guan, M.; Pan, W.; Zhang, J.; Hao, Q.; Cheng, J.; Zheng, X. Synchronous generator emulation control strategy for voltage source converter (VSC) stations. *IEEE Trans. Power Syst.* **2015**, *30*, 3093–3101. [CrossRef]

16. Yuko, H.; Kazushige, S.; Kenichi, S.; Toshifumi, I. Analysis of resonance in microgrids and effects of system frequency stabilization using a virtual synchronous generator. *IEEE J. Emerg. Sel. Top. Power Electron.* **2016**, *4*, 1287–1298.

17. Shi, R.L.; Zhang, X.; Liu, F.; Xu, H.; Hu, C.; Ni, H.; Yu, Y. A differential feedforward control of output current for high performance virtual synchronous generator. In Proceedings of the IEEE 2nd International Future Energy Electronics Conference (IFEEC), Taipei, Taiwan, 1–4 November 2015.

18. Liu, J.; Miura, Y.; Bevrani, H.; Ise, T. Enhanced virtual synchronous generator control for parallel inverters in microgrids. *IEEE Trans. Smart Grid* **2016**, *8*, 2268–2277. [CrossRef]

19. Hirase, Y.; Noro, O.; Sugimoto, K.; Sakimoto, K.; Shindo, Y.; Osaka, T.I. Effects and analysis of suppressing frequency fluctuations in microgrids using virtual synchronous generator control. In Proceedings of the 41st Annual Conference of the IEEE Industrial Electronics Society, Yokohama, Japan, 9–12 November 2015.

20. Shi, R.L.; Zhang, X.; Liu, F.; Xu, H.; Hu, C.; Yu, Y.; Ni, H. Research on power compensation strategy for diesel generator system based on virtual synchronous generator. In Proceedings of the IEEE 8th International Power Electronics and Motion Control Conference (IPEMC-ECCE Asia), Hefei, China, 22–26 May 2016.

21. Morren, J.; de Haan, S.; Kling, W.; Ferreira, J. Wind turbines emulating inertia and supporting primary frequency control. *IEEE Trans. Power Syst.* **2006**, *21*, 433–434. [CrossRef]

22. Xue, Y.; Tai, N. Review of contribution to frequency control through variable speed wind turbine. *Renew. Energy* **2011**, *36*, 1671–1677.

23. Alipoor, J.; Miura, Y.; Ise, T. Power system stabilization using virtual synchronous generator with alternating moment of inertia. *IEEE J. Emerg. Sel. Top. Power Electron.* **2015**, *3*, 451–458. [CrossRef]

24. Alipoor, J.; Miura, Y.; Ise, T. Distributed generation grid integration using virtual synchronous generator with adoptive virtual inertia. In Proceedings of the IEEE Energy Conversion Congress and Exposition, Denver, CO, USA, 15–19 September 2013.

25. Krishnamurthy, S.; Jahns, T.; Lasseter, R. The operation of diesel gensets in a certs microgrid. In Proceedings of the IEEE Power and Energy Society General Meeting—Conversion and Delivery of Electrical Energy in the 21st Century, Pittsburgh, PA, USA, 20–24 July 2008.

26. Shi, R.L.; Zhang, X.; Chao, H.; Xu, H.; Gu, J.; Cao, W. Self-tuning virtual synchronous generator control for improving frequency stability in autonomous photovoltaic-diesel microgrids. *J. Mod. Power Syst. Clean Energy* **2018**, *6*, 482–494. [CrossRef]

energies

MDPI

Article

Coordinated Control for Large-Scale Wind Farms with LCC-HVDC Integration

Xiuqiang He [1], Hua Geng [1,*], Geng Yang [1] and Xin Zou [2]

[1]   Department of Automation, Tsinghua University, Beijing 100084, China;
      he-xq16@mails.tsinghua.edu.cn (X.H.); yanggeng@tsinghua.edu.cn (G.Y.)
[2]   State Power Economic Research Institute, State Grid Corporation of China, Beijing 102209, China;
      zouxin@chinasperi.sgcc.com.cn
*   Correspondence: genghua@tsinghua.edu.cn; Tel.: +86-10-6277-0559

Received: 30 July 2018; Accepted: 21 August 2018; Published: 23 August 2018

Abstract: Wind farms (WFs) controlled with conventional vector control (VC) algorithms cannot be directly integrated to the power grid through line commutated rectifier (LCR)-based high voltage direct current (HVDC) transmission due to the lack of voltage support at its sending-end bus. This paper proposes a novel coordinated control scheme for WFs with LCC-HVDC integration. The scheme comprises two key sub-control loops, referred to as the reactive power-based frequency (*Q-f*) control loop and the active power-based voltage (*P-V*) control loop, respectively. The *Q-f* control, applied to the voltage sources inverters in the WFs, maintains the system frequency and compensates the reactive power for the LCR of HVDC, whereas the *P-V* control, applied to the LCR, maintains the sending-end bus voltage and achieves the active power balance of the system. Phase-plane analysis and small-signal analysis are performed to evaluate the stability of the system and facilitate the controller parameter design. Simulations performed on PSCAD/EMTDC verify the proposed control scheme.

Keywords: HVDC; line commutated converter; wind farm; frequency stability; frequency control; voltage stability; voltage control; vector control; voltage-source converter

---

1. Introduction

The power system is facing unprecedented technical challenges due to abundant large-scale renewable power plant integration [1–5]. In China, large-scale wind farms (WFs) are mainly built in remote areas of the northwest. As the local alternating current (AC) network is quite weak and the penetration level of wind power is extremely high, it is a critical technical issue to integrate and deliver large-scale wind power into the southeastern power grid. Emerging ultra HVDC (UHVDC) transmission technology is able to provide an available solution. To date, most of the UHVDC systems that are being planned or that are already built are of the line commutated converter (LCC) type, which are suitable for long-distance and large-capacity transmission, with advantages such as low expenditure and power loss [4]. Recently, a ±800 kV UHVDC with 10 GW capacity is being planned to deliver wind and solar power on the Tibetan Plateau into the eastern load center, and wind and solar power accounts for about 85% of the transmission capacity. It is quite difficult for the system to maintain stable operation when there is no traditional generating set that is available at the sending end of the UHVDC, due to some special factors, e.g., circuit faults, but only islanded WFs and/or photovoltaic power plants. Therefore, it is necessary to study the control scheme of large-scale WFs with LCC-HVDC integration as a technical reserve [4–7].

Vector control (VC) algorithms are commonly used for the control of wind energy conversion systems (WECSs). Conventional VC [8] is based on the orientation of the grid voltage vector, and thus a stiff grid is required to ensure the stability of the systems [9]. Unlike the voltage source converter in VSC-HVDC, the line commutated rectifier (LCR) in LCC-HVDC cannot actively generate the referenced three-phase voltage, essentially because of the application of semi-controlled switching devices, e.g., thyristors. Moreover, a steady and balanced commutation voltage is a prerequisite for the operation of the LCR. Consequently, under the condition that there is no available voltage support at the sending-end bus (SEB) of LCC-HVDC, the WFs with conventional VC algorithms cannot can be integrated by the LCC-HVDC directly [10–12].

A simple approach is to introduce a static synchronous compensator (STATCOM) on the SEB in order to provide the voltage support [10–12]. However, the extremely high reliability and large capability is required for the STATCOM, which results in high operating costs and power loss [13]. Without voltage support, the critical issue in the system is to guarantee the stability of the SEB voltage vector, including both the voltage stability and the frequency stability. Considering that the LCR is controllable in terms of active power, both the frequency and voltage stability issues can be addressed through the division of labor between the WF and the LCR [14–19], i.e., the voltage and the frequency are controlled by the WF and the LCR respectively, or conversely.

For the doubly-fed induction generator (DFIG)-based offshore WF at steady states, the stator voltage of the DFIG is the product of its stator flux and the SEB frequency [14]. Based on this fact, the earliest approach, where the stator flux and the frequency are controlled by the WF and the LCR respectively, is proposed in [14,15]. A similar approach can be found in [16], where the stator voltage and the frequency are controlled by the WF and the LCR respectively. In both of the approaches, the frequency is regulated by the active power of the WF, whereas the voltage is regulated by the reactive power of the WF. Actually, there is a substantial amount of capacitive compensation on the SEB of the LCC-HVDC or diode-based HVDC, leading to a strong coupling between the bus voltage and the active power balance, and also between the frequency and the reactive power balance [17,20–22]. Consequently, references [17,20–22] develop a novel control concept, where the frequency is regulated by the reactive power, whereas the voltage is regulated by the active power. There is another coordination approach developed in [18,19] where the voltage is controlled by both the WF and the LCR. As a result, the strong coupling between the active and reactive power control loops occurs, which would affect the dynamic performance of the system. Moreover, the frequency stability was not addressed in the approach.

In the aforementioned approaches, the control algorithms of the WFs are based on the conventional VC structure where phase-locked loops (PLLs) are employed to detect the phase-angle of the stator voltage. It is reported that PLLs play an important role in the system dynamics, and the system stability involving PLLs are quite complicated [9], especially when WFs are connected to the weak, or even isolated grids [23,24]. There are some intensive studies regarding the PLL-less DFIG control algorithms for standalone applications, referred to as the indirect self-orientated vector control (ISOVC) [25–28]. In the ISOVC, the phase-angle, adopted in the coordinate transformation between *abc* and *dq* reference frames, is derived from a free running integral of the rated synchronous speed ω_0 instead of the PLL. It is worth noting that the supplementary indirect orientation control is realized through modifying the original active power loop, and thus the auxiliary torque and pitch angle control is required to regulate the active power [25–28]. Since the probability of frequency instability due to the dynamic characteristics of PLLs can be completely avoided in the ISOVC, it is applied to control standalone DFIGs with LCC-HVDC integration in [29]. In order to be employed in multi-machine scenarios, additional active power droop loop should be introduced into ω_0 to achieve synchronization and power sharing among multiple machines [29,30]. Unfortunately, with such droop scheme, the WF cannot always track its maximum power point with sacrificed economic benefits.

In this paper, a novel scheme with respect to the division of labor between the WF and the LCR is proposed. On one hand, considering the coupling relationship between the frequency and the reactive power balance [17–20], a novel indirect orientation control based on the reactive power loop instead of the active power loop [25–28] is developed. In the scheme, reactive power droop is employed for synchronization and reactive power sharing among multiple machines. Therefore, it would not affect the active power tracking of the WF. On the other hand, the control objective of the LCR is to maintain the SEB voltage stability. Actually, since there is a strong coupling between the voltage and the active power balance, not only can the voltage stability be addressed, but also the WF is able to capture the maximum active power with the proposed scheme. The proposed scheme comprises two key sub-control loops. One is the reactive power-based frequency (*Q-f*) control loop for the voltage source inverters (VSIs) in the WF, where a novel ISOVC is developed to maintain the SEB frequency and compensate reactive power for the LCR. The other is the active power-based voltage (*P-V*) control loop for the LCR, through which the SEB voltage is controlled to achieve the active power balance.

The rest of the paper is organized as follows. Section 2 depicts the mathematical models, and explains the relationship between the voltage and the active power balance, and also between the frequency and the reactive power balance. Section 3 proposes the *Q-f* and *P-V* control loops after analyzing the operational principles. Section 4 demonstrates the stability of the proposed control scheme and designs the controller parameters. Section 5 shows the simulation results and verifies the feasibility of the coordinated control scheme. Section 6 concludes this paper.

2. System Modeling

The topology of the studied system is shown in Figure 1. This study takes permanent-magnet synchronous generator (PMSG)-based WECSs as an example to study the coordination between WFs and LCC-HVDC. The proposed control scheme can be easily extended to doubly-fed induction generator (DFIG)-based WECSs. It should be noticed that the *Q-f* control is applied in grid-side VSIs of PMSG-based WECSs, whereas it is applied in rotor-side converters of DFIG-based WECSs. For simplicity, the PMSG-based WECS can be equivalent to a voltage source inverter (VSI) in parallel with a direct current (DC) capacitor and a controlled DC current source [31]. In order to supply energy for the system startup, batteries are installed at the DC bus of several units (no need for all units). Especially, for the DFIG-based WECSs, rotor excitation of DFIGs should be provided initially in the startup process, which can be accomplished through the rotor-side converters powered by the batteries. The WECSs are connected to the SEB of HVDC. The rectifier is of the LCC type, and thus a substantial amount of AC filters are configured at the SEB to mitigate the current harmonics, and meanwhile they provide reactive power compensation, which can be equivalent to a capacitor bank C_f at the fundamental frequency.

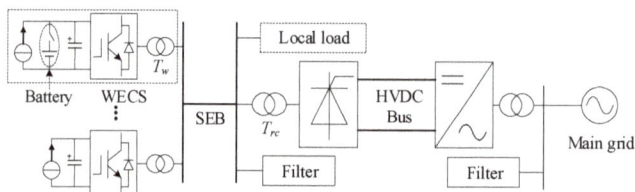

Figure 1. System topology.

For a clear description of the system model, it can be divided into three subsystems: the WECS subsystem, the SEB subsystem and the HVDC subsystem. To capture the fundamental power dynamic characteristics of the system, the switching function models [10,11] of the converters are employed. Moreover, the inverter of HVDC can be equivalent to a DC voltage source [14,15] since it is subjected to a constant-voltage control and has little effect on the other subsystems under normal operations. Figure 2 depicts the equivalent circuit of the whole system, where the significations of the electrical variables and parameters are self-explanatory.

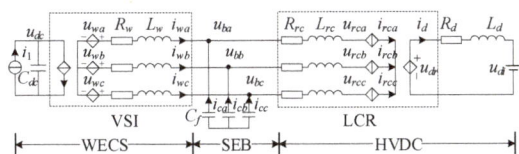

Figure 2. Equivalent circuit, where the wind farm (WF) is represented by a single wind energy conversion systems (WECS) just for a simplified system model. Note that the proposed control scheme is applicable for multiple WECSs (see the next section for more details).

2.1. WECS Subsystem Model

In the rated synchronous reference frame (RSEF) with the rated frequency ω_0, the WECS model can be written as:

$$\begin{cases} L_w di_{wd}/(\omega_b dt) = -R_w i_{wd} - u_{bd} + m_d u_{dc} + \omega_0 L_w i_{wq} \\ L_w di_{wq}/(\omega_b dt) = -R_w i_{wq} - u_{bq} + m_q u_{dc} - \omega_0 L_w i_{wd} \end{cases} \tag{1}$$

$$C_{dc} du_{dc}/(\omega_b dt) = i_1 - (m_d i_{wd} + m_q i_{wq}) \tag{2}$$

where the subscripts d and q represent the variables transformed from the three-phase abc reference frame to the dq RSEF, and ω_b is the base frequency, similarly hereinafter.

2.2. SEB Subsystem Model

Similarly, the SEB model in the RSEF can be written as:

$$\begin{cases} C_f du_{bd}/(\omega_b dt) = i_{wd} - i_{rcd} + \omega_0 C_f u_{bq} \\ C_f du_{bq}/(\omega_b dt) = i_{wq} - i_{rcq} - \omega_0 C_f u_{bd} \end{cases} \tag{3}$$

Let the amplitude and the phase-angle of the SEB voltage be:

$$U_{bm} = \sqrt{u_{bd}^2 + u_{bq}^2}$$
$$\phi = \arctan\left(u_{bq}/u_{bd}\right), \phi \in [0, 2\pi) \tag{4}$$

It can be obtained in the polar coordinate system that [19,20]:

$$0.5 dU_{bm}^2/(\omega_b dt) = (P_w - P_{rc})/C_f \tag{5}$$

$$\omega_0 + d\phi/(\omega_b dt) = \omega_1 = (-Q_w + Q_{rc})/\left(C_f U_{bm}^2\right) \tag{6}$$

where $P_w = u_{bd} i_{wd} + u_{bq} i_{wq}$ and $P_{rc} = u_{bd} i_{rcd} + u_{bq} i_{rcq}$ are the active powers from the WECS, and they are absorbed by the HVDC respectively. Also, $Q_w = -u_{bd} i_{wq} + u_{bq} i_{wd}$ and $Q_{rc} = -u_{bd} i_{rcq} + u_{bq} i_{rcd}$ are the reactive powers from the WECS and they are absorbed by the LCR respectively, and ω_1 is the SEB frequency.

Equations (5) and (6) lay the foundations for this study. From the perspective of the filter capacitor parallel branch, the WECS can be seen as a controlled power source (P_w and Q_w), while the LCR can be seen as a controlled power load (P_{rc} and Q_{rc}). Given that the WECS and the HVDC are interconnected by the filter capacitor, two significant results can be drawn from the filer capacitor point of view. (1) The voltage amplitude U_{bm} is highly coupled with the active power ($P_w - P_{rc}$); (2) The phase-angle ϕ (or the frequency ω_1) is highly coupled with the reactive power ($Q_w - Q_{rc}$) [20]. A physical mechanism explanation is given as follows.

It is known that the total instantaneous power of the three-phase balanced capacitor branch equals its active power (which is zero) at steady states, and the power exchanged within the three-phase capacitors charges/discharges the capacitors. As a result, the three-phase capacitor circuit as a whole exhibits a certain reactive power. However, this conclusion becomes invalid during dynamic processes. For Equation (5), if there is an active power deviation, e.g., $P_w - P_{rc} > 0$, this indicates that the active power generated by the WECS is larger than that absorbed by the LCR, then the extra active power will charge the capacitors. Consequently, the instantaneous current amplitude will increase, which leads the instantaneous voltage amplitude to increase too. Similarly, if there is a reactive power deviation for Equation (6), e.g., $(-Q_w + Q_{rc}) > 0$, which indicates that the reactive power generated by the WECS is smaller than that absorbed by the LCR, then the voltage phase-angle (i.e., the instantaneous frequency) will increase, assuming that the voltage amplitude keeps unchanged. As a consequence, the capacitive reactance will decrease due to the frequency increase, and therefore the capacitor will generate more reactive power to try to balance the reactive power. In fact, the capacitive parallel branch is in a dual relationship with the inductive series branch in traditional power systems, and thus it is not difficult to understand the foregoing coupling relationship. Moreover, it can be also found that a smaller C_f (the absolute minimum filter guarantees $C_f > 0$) can result in a stronger coupling.

2.3. HVDC Subsystem Model

In the RSEF, the HVDC model can be written as [11,12]:

$$\begin{cases} L_{rc} \, di_{rcd}/(\omega_b dt) = -R_{rc}i_{rcd} + u_{bd} - u_{rcd} + \omega_0 L_{rc}i_{rcq} \\ L_{rc} di_{rcq}/(\omega_b dt) = -R_{rc}i_{rcq} + u_{bq} - u_{rcq} - \omega_0 L_{rc}i_{rcd} \end{cases} \tag{7}$$

$$\begin{cases} u_{rcd} = U_{rcm}\left(i_{rcd} \sin \alpha + i_{rcq} \cos \alpha\right) \\ u_{rcq} = U_{rcm}\left(i_{rcd} \cos \alpha - i_{rcq} \sin \alpha\right) \end{cases} \tag{8}$$

$$u_{dr} = U_{rcm} \cos \alpha \tag{9}$$

$$L_d di_d /(\omega_b dt) = u_{dr} - u_{di} - R_d i_d \tag{10}$$

where α is the firing angle, and U_{rcm} is the voltage amplitude in the rectifier bridge side. Note that the variables are in the per-unit system, and thus the rectifier coefficient is eliminated. Also, both the ($12k \pm 1$) order harmonics in ac side and the $12k$ order harmonics in DC side are neglected in the typical 12-plus HVDC model, since the events of major concern are the fundamental power conversion rather than high frequency dynamic.

3. Coordinated Control Scheme

Prior to describing the proposed coordinated control scheme, the control requirements should be emphasized first.

(1) Voltage control: a stable voltage can offer voltage support for the WECSs, as well as the commutation voltage for the LCR.
(2) Frequency control: the frequency stability should be maintained so that multiple WECSs are able to operate synchronously.
(3) Active power balance: considering that the wind conditions are not controlled, the active power generated from the WFs should be equal to that which is transmitted into the HVDC in real time.
(4) Reactive power balance: considering that the reactive power compensation capability of AC filters is discontinuous, the WECSs should be able to compensate and share the insufficient or excessive reactive power automatically.

The proposed coordinated control scheme is able to achieve the requirements. The *Q-f* control applied into the wind farm meets the requirements (2) and (4), whereas the *P-V* control applied into the HVDC rectifier meets the requirement (1) and (3). Thus, the wind farm and the HVDC cooperate with each other to achieve system stability.

Note that the resynchronization capability is also of much significance for the system uninterrupted operation in the case of a fault. Under fault conditions, the back-end converters can be controlled to supply zero power temporarily, and then the batteries can be utilized again to help generate the SEB voltage after the fault is cleared. More technical details will be given in future work.

3.1. Q-f Control of WECSs

In contrast to the ISOVC in [25–28], where the active power loop is adopted to achieve indirect orientation and synchronization, a novel ISOVC will be developed here. Since that the relationship between the frequency and reactive power, as shown in Equation (6), the reactive power loop is adopted to achieve Q-f control. As depicted in Figure 3, two key modifications are made in the Q-f control compared with the conventional VC with the unity power factor. One is that the self-defined phase-angle is:

$$\angle U = \int \omega_0 + \angle_0 \tag{11}$$

where $\angle_0$ is an initial value, determined by the final value of the phase-angle at the end of system startup. The other is that the control object of the reactive power loop is regulating the q-axis voltage u_{bq} instead of the reactive current i_{wq} to zero. As a consequence, not only is the frequency ω_1 clamped when $u_{bq} = 0$ since the actual phase-angle is consistent with the self-defined one, but the reactive current command $i_{wq}*$ can also be regulated automatically so that the WECS is able to compensate reactive power for the LCR. Note that the active power control is still based on the maximum power point tracking (MPPT) control, as shown in Figure 3b. The operational principle of the Q-f control is described as follows.

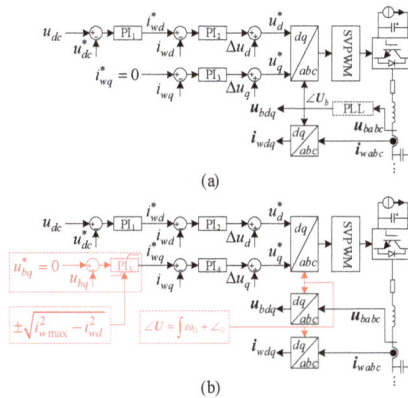

(a)

(b)

Figure 3. (**a**) Conventional vector control (VC) versus (**b**) proposed reactive power-based frequency (Q-f) control for voltage source inverters (VSIs) of WECSs. In actual practice, the controlled variable u_{bq} can be replaced by the local voltage information instead of sending-end bus (SEB) voltage information. Since the critical information is the voltage-phase angle rather than the voltage amplitude, the spatial distribution feature of the voltage amplitude has litter influences on the Q-f control performance.

Assuming that the system is in a steady state, if the frequency ω_1 suddenly starts to increase due to a disturbance, e.g., a sudden increase of the reactive power Q_{rc}, the voltage vector U_b and current vector I_w under conventional VC is shown in Figure 4a. It can be observed that the controller tracks the frequency change and cannot output the reactive power to regulate the frequency, which further leads to the instability of frequency. In Figure 4b, the self-oriented control (11) is adopted, but the reactive power loop still regulates i_{wq} to zero. Under this condition, U_b rotates counterclockwise an angle ϕ due to the increase of frequency. Thus, although the WECS outputs reactive power, U_b no longer coincides with the q-axis because of a lack of synchronization control. Thereafter, the d- and q-axis controls are no longer decoupled, and moreover, multiple WECSs may become asynchronous due to the accumulation of the angle errors. In Figure 4c, the issue is completely addressed where the control object is $u_{bq} = 0$ instead of $i_{wq} = 0$. In other words, the phase-angle is indirectly controlled to follow the angle generated by the rated frequency ω_0, and the controller output signal is exactly the reactive current reference. Consequently, the third proportional-integral (PI$_3$) regulator is able to produce an exact reactive current reference $i_{wq}{}^*$ under the condition that $u_{bq} = 0$.

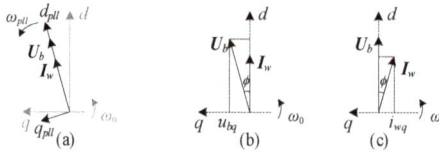

Figure 4. Vector diagrams in the rated synchronous reference frame (RSRF) under different control algorithms: (**a**) Conventional VC with $i_{wq} = 0$; (**b**) Self-oriented control with $i_{wq} = 0$; (**c**) Proposed Q-f control with $u_{bq} = 0$.

It should be noted that the PI$_3$ regulator cannot work well in multi-WECS scenarios due to a reactive current circulation among the WECSs without a sharing scheme. To this end, a simple approach is to improve the PI-type regulator into a P-type one, resulting in a droop characteristic. Thus, when a multi-machine system is subjected to the Q-f droop control, the reactive power can be compensated and shared automatically, and both the frequency stability and the synchronization stability can be realized. With the P-type control, the voltage vector will no longer coincide with the d-axis. By defining an appropriate range of the included angle between them, and thereby setting an appropriate proportional coefficient, it is doable to ensure that the included angle is small enough and close to zero at steady states.

According to the P-type control, for one WECS$_j$, it can be obtained that:

$$i_{wqj} = -k_{pj}u_{bqj} \tag{12}$$

where u_{bqj} can be considered to be approximately the same for different units. For one thing, in actual practice, $u_{bqj} = 0$ at the time when WECS$_j$ switches from the pre-synchronization stage to the connection to the sending-end grid by means of phase-locked loops. For another, after the connection to the sending-end grid, the phase-locked loops are withdrawn. Then, the phase-angle difference between two units during normal operating conditions are also eliminated by their initial synchronous reference frames.

On the basis of Equation (12), the reactive power can be written as:

$$Q_{wj} = -u_{bdj}i_{wqj} + u_{bqj}i_{wdj} = \left(k_{pj}u_{bdj} + i_{wdj}\right)u_{bqj} = k'_{pj}u_{bqj} \tag{13}$$

Figure 5 shows the droop curves of the reactive currents and the reactive powers with respect to the d-axis voltage. It can be observed in Figure 5a that a large proportional coefficient is able to result in a large shared reactive current. In Figure 5b, the sharing coefficient is given by Equation (13). Since the voltage vector and the d-axis are not strictly coincident, the reactive power is inevitably related to the active current i_{wdj}. Even so, if a quite large k_{pj} is taken, then the voltage vector and the d-axis will substantially coincide. In this point, it is known from Equation (13) that the reactive power contributed by the active current i_{wdj} is relatively small. In short, the proportional coefficient of each unit can be calculated and designed based on Equation (13). In practical applications, the optimization control and the quantitative allocation of reactive power can be achieved considering the unit capacity limit. For example, a large proportional coefficient can be set for the unit with a small active power output so that it can share more reactive power.

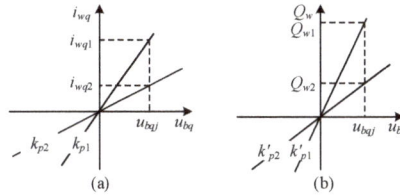

Figure 5. Reactive current and reactive power sharing relationship with the P-type droop control. (a) Reactive current sharing relationship; (b) reactive power sharing relationship.

In order to maintain the active power benefit of WFs, the active power priority principle can be adopted. As shown in Figure 3, according to the real-time active current i_{wd}, the reactive current i_{wq} is limited as follows:

$$i_{wq\lim} = \pm\sqrt{i_{wmax}^2 - i_{wd}^2} \tag{14}$$

Equation (14) defines the reactive power margin. There is no steady-state equilibrium point of reactive power, assuming that the reactive power demand exceeds the reactive power margin. Considering that the reactive power demand can be adjusted by the centralized reactive power compensation device, and that a WF has the minimum reactive power compensation capability, this study recommends the following reactive power compensation scheme: (1) For the entire WF, according to its rated capacity and maximum active power output to determine the minimum reactive power compensation capability. When the real-time reactive power output becomes larger than the minimum compensation capacity, the centralized reactive power compensation should be adjusted in time, such as the conventional filter or the high-performance static var compensator (SVC) or STATCOM. Consequently, the reactive power demand becomes smaller, and within the minimum compensation capability. (2) For each unit, according to its real-time active current, adjust the limitation of the reactive current according to Equation (14).

3.2. P-V Control of LCR

When the active power P_w from the WECS increases, it has been known that U_{bm} increases according to Equation (5) assuming that P_{rc} remains constant. Actually, if U_{bm} increases, P_{rc} will increase too. However, there is no doubt that the steady-state U_{bm} will become larger from the perspective of the whole circuit if the firing angle remains unchanged. Only if the LCR controller reduces the firing angle α, leading to more absorbed active power P_{rc} by the LCR, can U_{bm} return back its reference value. A detailed analysis is performed as follows.

Based on the conclusions in [5], the time constant of the rectifier currents is quite small, about tens of milliseconds, under the constant-voltage control of the inverter of HVDC. Thus, the current transients in Equations (7) and (10) can be ignored while analyzing the active power balance, which gives rise to:

$$\begin{aligned} P_{eq} &= U_{bm}U_{rcm}\sin\delta/(\omega_0 L_{rc}) \\ Q_{eq} &= U_{bm}U_{rcm}\cos\delta/(\omega_0 L_{rc}) - U_{rcm}^2/(\omega_0 L_{rc}) \end{aligned} \tag{15}$$

where P_{eq} and Q_{eq} are the active and reactive power of rectifier bridge, and δ is the phase-angle difference between U_b and U_{rc}, and it can be seen as a power angle. Note that the internal resistor R_{rc} is ignored in Equation (15). Given that the power factor angle of the rectifier bridge (excluding R_{rc} and L_{rc}) is α, i.e.,:

$$Q_{eq} = P_{eq}\tan\alpha \tag{16}$$

Combining Equations (9), (10), (15), and (16), it can be obtained that:

$$U_{bm}\cos(\delta+\alpha) = u_{dr} = u_{di} + i_d R_d \approx u_{di} \qquad (17)$$

Furthermore, considering the effect of the leakage inductor L_{rc} on the power factor angle of the LCR (including T_{rc}), it exists as:

$$u_{dr} = U_{bm}\cos\alpha - i_d R_c \qquad (18)$$

where R_c is the equivalent commutation resistor with an actual value $6/\pi\omega_0 L_{rc}$. Substituting Equation (18) into Equations (9) and (10) yields that:

$$U_{bm}\cos\alpha = u_{di} + i_d(R_c + R_d) \qquad (19)$$

It can be assumed that both U_{bm} and u_{di} remain constant under the controls of both the rectifier and inverter, and thus $\cos(\delta+\alpha)$ remains constant, according to Equation (17). Actually, when P_w increases, the power angle δ increases, whereas the firing angle α decreases along with the increase of i_d in Equation (19). Therefore, $\delta + \alpha$ remains approximately unchanged. Since that $\delta + \alpha$ is the power factor angle between U_b and I_{rc}:

$$\begin{aligned} P_{rc} &= U_{bm}I_{rcm}\cos(\delta+\alpha) \\ Q_{rc} &= U_{bm}I_{rcm}\sin(\delta+\alpha) \end{aligned} \qquad (20)$$

Let U_b be oriented to the d-axis of the rated synchronous reference frame (RSRF) (consistent with the orientation of the WECS), and the vector diagrams of the SEB is depicted in Figure 6. In Figure 6a, the active current i_{wd} is smaller, and it can be assumed that the reactive current I_c from the filters can be supplied to the rectifier exactly. In Figure 6b, however, i_{wd} becomes larger, resulting in more required reactive power for the LCR. Assuming that I_c remains constant due to a time-delay to adjust the filters, the voltage amplitude U_{bm}, and the phase-angle ϕ of U_b are controlled by the LCR and the WECS respectively, and then the WECS can output the required reactive current automatically. Thus, I_c, together with i_{wq}, is able to provide the reactive current of the LCR. Moreover, the power angle increases, whereas the firing angle decreases, and U_b remains constant.

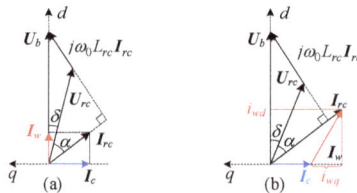

Figure 6. Vector diagrams under the condition that U_b remains constant. (**a**) A smaller active current i_{wd} (or P_w); (**b**) a larger active current i_{wd}.

After analyzing the power characteristics of the system, the *P-V* control loop can be designed as follows. According to Equation (19), there is a linear relationship between $U_{bm}\cos\alpha$ and i_d. In fact, α changes within a narrow range of around 20° [14], and therefore, it can be approximately considered that $\cos\alpha$ remains unchanged, leading to an approximate linear relationship between U_{bm} and i_d. Consequently, a linear regulator, such as PI, can be employed to control U_{bm}, as shown in Figure 7, and its output is the DC current reference $i_d{}^*$. In the inner current loop, the typical regulator is applied. Note that the optional compensation can be performed in the outer loop, so as to obtain the same control gain at different operating points.

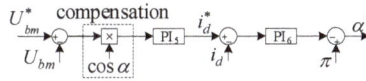

Figure 7. Proposed power based voltage (P-V) control for LCR.

4. Stability Analysis and Parameter Design

4.1. Stability Analysis of Q-f Control

While analyzing the stability of the Q-f control of the WECS subsystem, the voltage amplitude U_{bm} can be assumed to be constant, and both the active current i_{wd} and the reactive power Q_{rc} are seen as external disturbances. Moreover, the current dynamics can be neglected since they are generally much faster than those of power. Considering only the outer loop of the Q-f control, it can be obtained that:

$$i_{wq} = (k_{p3} + k_{i3}/s)(0 - U_{bm}\sin\phi) \tag{21}$$

Let $i_{wq} = i'_{wq} + k_{p3}(0 - U_{bm}\sin\phi)$, and rewrite Equations (6) and (21) as:

$$d\phi/dt = \omega_b[U_{bm}(i'_{wq} - k_{p3}U_{bm}\sin\phi)\cos\phi \\ -U_{bm}i_{wd}\sin\phi + Q_{rc}]/(C_fU_{bm}^2) - \omega_b\omega_0 \tag{22}$$

$$di'_{wq}/dt = -k_{i3}U_{bm}\sin\phi \tag{23}$$

From Equations (22) and (23), it can be observed that there is a coupling relationship between the state variables ϕ and i'_{wq}, which determines the reactive power-frequency dynamic characteristics.

The phase-plane analysis [32] can be performed to demonstrate the stability of the simplified second-order nonlinear system consisting of Equations (22) and (23). Clearly, the equilibrium point of the system is:

$$\begin{cases} \phi = 0 \\ i'_{wq} = (\omega_0 C_f U_{bm}^2 - Q_{rc})/U_{bm} \end{cases} \tag{24}$$

Let $a = \omega_b(k_{p3}U_{bm} + i_{wd})/(C_fU_{bm})$, and $b = k_{i3}\omega_b/C_f$. The linearized system at Equation (24) is $\Delta\dot{x} = A\Delta x$ with $\Delta x = [\Delta\phi, \Delta i'_{wq}]$ and:

$$A = \begin{bmatrix} -a & b/(k_{i3}U_{bm}) \\ -k_{i3}U_{bm} & 0 \end{bmatrix} \tag{25}$$

The characteristic equation is $\lambda^2 + a\lambda + b = 0$. Clearly, the system is small-signal stable, since $a > 0$ and $b > 0$. If $a^2 - 4b < 0$, Equation (24) will be a stable focus, otherwise it is a stable node. In the former case, the motion near the equilibrium point will converge in the form of oscillations. While in the latter case, there is no oscillation during the convergence and an asymptote exists around the equilibrium point:

$$\Delta i'_{wq} = -k_{i3}U_{bm}/\lambda_1\Delta\phi \tag{26}$$

where λ_1 is the eigenvalue with a smaller modulus.

Taking a concrete case as an example: $U_{bm} = 1.0$ pu, $i_{wd} = 0.8$ pu, $Q_{rc} = C_f = 0.21$ pu in an initial state, and they remain unchanged in the following convergence process. Also, set $k_{p3} = 0.6$, $k_{i3} = 50$. Since that the reactive power from C_f is exactly supplied to Q_{rc}, since $Q_{rc} = C_f$, $Q_w = 0$, and the equilibrium point is the origin. As shown in Figure 8, the equilibrium points of Equation (22) and those of Equation (23) form the two equilibrium curves respectively in the $\phi - i'_{wq}$ phase plane, and the intersection, i.e., the origin O, is the equilibrium point of the system. In Figure 8, the range of the state variable i'_{wq} is $[i'_{wq\min}, i'_{wq\max}]$.

$$\begin{aligned} i'_{wq\max} &= i_{wq\max} + k_{p3}U_{bm}\sin\phi = \sqrt{i^2_{w\max} - i_{wd}} + k_{p3}U_{bm}\sin\phi \\ i'_{wq\min} &= i_{wq\min} + k_{p3}U_{bm}\sin\phi = -\sqrt{i^2_{w\max} - i_{wd}} + k_{p3}U_{bm}\sin\phi \end{aligned} \tag{27}$$

where $i_{w\max}$ is the maximum current of the WECS. Moreover, the phase-angle ϕ should be in the range $(-\pi/2, \pi/2)$ so as to ensure the negative feedback property of the controller.

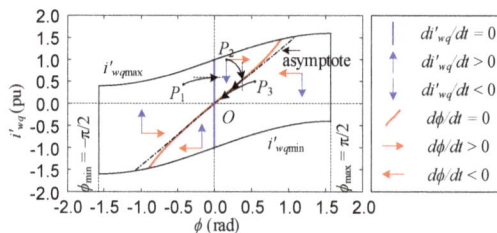

Figure 8. Phase plane $\phi - i'_{wq}$ with the Q-f control.

In Figure 8, the domain $(-\pi/2, \pi/2) \times [i'_{wq\min}, i'_{wq\max}]$ is divided into four sections. In each section, the horizontal and vertical arrows indicate the directions of motions of Equations (22) and (23) respectively, and the actual direction of the state trajectory is the synthesis direction. Taking the initial point P_1 as an example, as shown by the black arrow, the trajectory traverses the blue curve vertically and then enters the section where P_2 is located. Then, it moves to the lower right, and traverses the red curve in a vertical direction. Thereafter, the state trajectory will move along the asymptote to the steady state point. Similar cases occur in other sections. Therefore, it can be concluded that the system is locally asymptotically stable in the domain.

It should also be assumed that the initial point is I with the position $(-0.1, 0.1)$, and the convergence process of the second-order simplified system is illustrated in Figure 9. From Figure 9a, it can be observed that the system converges along the asymptote, since the equilibrium point is a stable node. In Figure 9b, the equilibrium point become a stable focus due to a large k_{i3}. Thus, although the convergence speed become large, there is an overshoot in the motion near the stable focus. Therefore, it can be concluded that a large k_{i3} can facilitate the system converge speed, causing both overshoot and oscillation. A trade-off should be considered to design an appropriate value for the control parameter k_{i3}. Similar work can be made to further study the effects of other parameters on the system dynamic behaviors.

It is noteworthy that, when i'_{wq} reaches the limitation $i'_{wq\min}$ or $i'_{wq\max}$, the system state moves along the boundaries and it can still converge once the state reaches another section across the blue curve as long as the equilibrium point is in the allowable domain. In Equation (27), the range of i'_{wq} is related to i_{wd}, i.e., the active power P_w. While in Equation (24), the equilibrium point is related to the required reactive power Q_w. The different equilibrium curves (or points) and domains under different P_w and Q_w are depicted in Figure 10. In particular, the equilibrium point P_1 (P_1') are beyond the domain, when both P_w and Q_w are quite large, such as 1.0 and 0.9 (-0.9). The problem must be avoided in practice. To this end, a proper capacity of filters could be designed to result in a smaller Q_w.

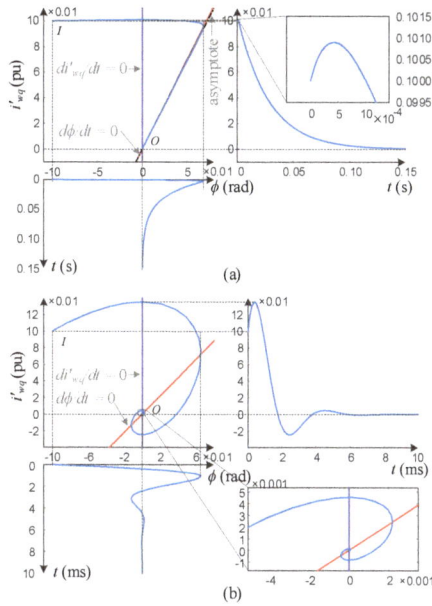

Figure 9. Convergence process of the second-order simplified system composed of Equations (22) and (23), from the initial point *I*. (**a**) $k_{p3} = 0.6$, $k_{i3} = 50$; (**b**) $k_{p3} = 0.3$, $k_{i3} = 2000$.

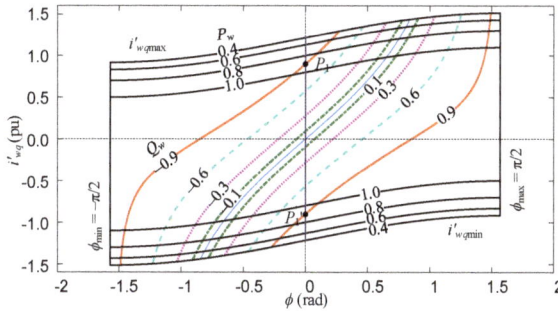

Figure 10. Different equilibrium curves and domains under different P_w and Q_w.

4.2. Stability Analysis of P-V Control

A simplified second-order equation of the HVDC subsystem can be obtained when the current transients are ignored.

$$i_d \approx (k_{p5} + k_{i5}/s)(U_{bm}^* - U_{bm}) \tag{28}$$

$$dU_{bm}^2/dt \approx 2\omega_b \left(P_w - u_{di}i_d - R_d i_d^2 \right)/C_f \tag{29}$$

According to Equations (28) and (29), when P_w increases, the voltage amplitude U_{bm} will increase, and thus the control loop Equation (28) will adjust the DC current reference. Once the firing angle regulated by the inner loop decreases, the actual DC current i_d will increase. According to Equation (29), U_{bm} will stop increasing and start to decrease, and finally the system will reach a new steady state.

4.3. Small-Signal Analysis and Parameters Design

Taking one of typical operating points, i.e., $P_w = 0.8$ pu and $Q_{rc} = 0.21$ pu; as an example, the small-signal model of the overall system can be established in the RSEF, as shown in Figure 11. For the inner loop regulator, PI_2 and PI_4, of the WECS, the parameters can be set as k_{p2} (k_{p4}) is 1.0, and k_{i2} (k_{i4}) is 10, leading to a closed-loop bandwidth about 167 Hz. Moreover, typical parameters such as $k_{p1} = 4.0$, $k_{i1} = 50$ can be set for the outer loop active power regulator PI_1, leading to a closed-loop bandwidth of about 20 Hz.

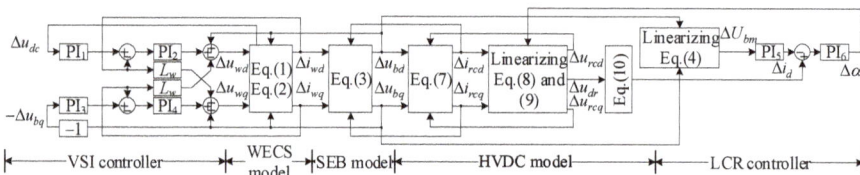

Figure 11. Small-signal model of the system.

In particular, attention should be paid to the parameters of the developed outer loop reactive power control. As shown in Figure 12a, when k_{i3} becomes larger, the damping of PI_3 increases. However, the loop will couple with the DC voltage when k_{i3} is too large, resulting in a pair of conjugate modes. It can be seen from Figure 12b that when k_{p3} increases, the damping of PI_3 decreases, which thereafter leads to the coupling. However, the coupling disappears as k_{p3} continues to increase. Finally, $k_{p3} \in [1.5, 2.0]$, and $k_{i3} \in [20, 140]$ can be taken. Within the tuned parameter ranges, the closed-loop bandwidth is about 40 Hz.

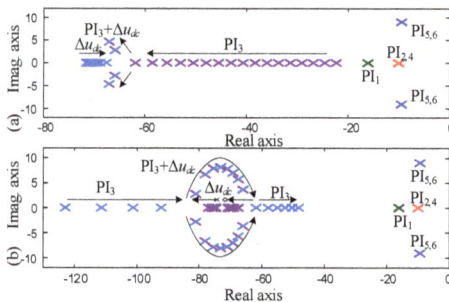

Figure 12. Root loci of the small-signal model. (a) $k_{p3} = 1.3$ and k_{i3} changes from 50 to 140; (b) $k_{i3} = 130$ and k_{p3} changes from 0.3 to 2.0.

A similar analysis can be performed to tune the parameters of the rectifier *P-V* controller. In Figure 13a, when k_{i5} becomes larger, the oscillation mode in $PI_{5,6}$ disappears, but the damping of PI_6 tends to decrease and the damping characteristics of both the voltage and current of the SEB are deteriorated. In Figure 13b, the larger k_{p5} is, the smaller the damping factor of $PI_{5,6}$ is. Finally, $k_{p5} \in [0.2, 0.5]$, $k_{i5} \in [40, 100]$ can be selected. Note that typical parameters such as $k_{p6} = 2.0$ and $k_{i6} = 20$ are employed in the inner loop. Within the suggested parameter ranges, the outer and inner closed-loop bandwidths are approximately 5 Hz and 110 Hz respectively.

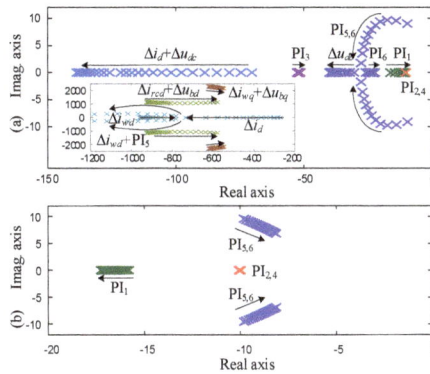

Figure 13. Root loci of the small-signal model. (**a**) k_{p5} = 0.4 and k_{i5} changes from 50 to 1000; (**b**) k_{i5} = 50 and k_{p5} changes from 0.1 to 2.0.

5. Simulation Results

Simulations are carried out on PSCAD/EMTDC to verify the proposed control scheme. The simulated system is shown in Figure 14, and the detailed parameters of the system are listed in Table 1. The employed monopole LCC-HVDC model is from the CIGRE benchmark model [33], both the rectifier and inverter of which are LCC-based. The rated capability of the system is 1000 MVA, and the reactive power capability of each set of AC 11/13 order harmonic filters is 50 MVar. The increase and decrease of the DC-side current of the WECS can simulate the changes of the input power of the back end.

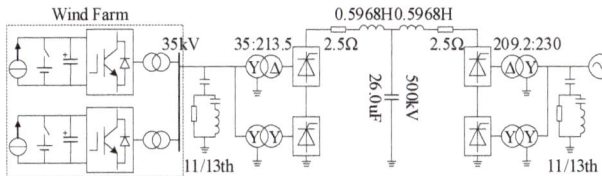

Figure 14. Simulated system.

Table 1. Parameters of the simulated system.

	C_{dc}	90,000 uF for 1.5 MVA capacity
Wind Energy Conversion System	R_w	0.001 pu
	L_w	0.3 pu
Sending-End Bus	C_f	0.05 pu for each set of filter
	R_{rc}	0.001 pu
High Voltage Direct Current System	L_{rc}	0.18 pu
	L_d	1.1936 H
	R_d	5 Ω

5.1. System Startup

The WF is equivalent to a single WECS, and the system startup process is shown in Figure 15. Before the system startup, the capacitor of the WECS is charged by the configured battery, and thereafter the three-phase AC voltage of the SEB can be generated by the WECS at 0–0.3 s, as shown in Figure 15a. At 0.3 s, the DC-side current of the WECS starts to increase, and then the HVDC is unblocked, with the

P-V control in the rectifier whereas the constant-voltage control in the inverter. Henceforth, the DC voltage of HVDC is generated gradually, as shown in Figure 15g. At 0.4 s, the battery configured at the DC bus of the WECS is withdrew, and the *Q-f* control in the WECS is switched on. Then, the active power continues increasing until to the rated point, as shown in Figure 15b,f. In Figure 15a, it can be observed that both the system frequency ω_1 and the voltage of the SEB U_{bm} remain stable and are maintained at 1.0 pu in the final steady state. The active power generated from the WECS can be delivered into the receiving end grid through the LCC-HVDC transmission. The AC filters are connected to the SEB gradually when the reactive power from the WECS Q_w is large than 0.1 pu (see Figure 15c), which can be regarded as the reactive power limit of the WECS. The reactive current i_{wq} of the WECS is regulated automatically, so as to compensate the reactive power for the LCR under the condition that $u_{bq} = 0$, as shown in Figure 15d,f. Moreover, Figure 15e indicates that the DC-side voltage of the WECS remains stable under the active power control of the WECS after the battery is withdrawn.

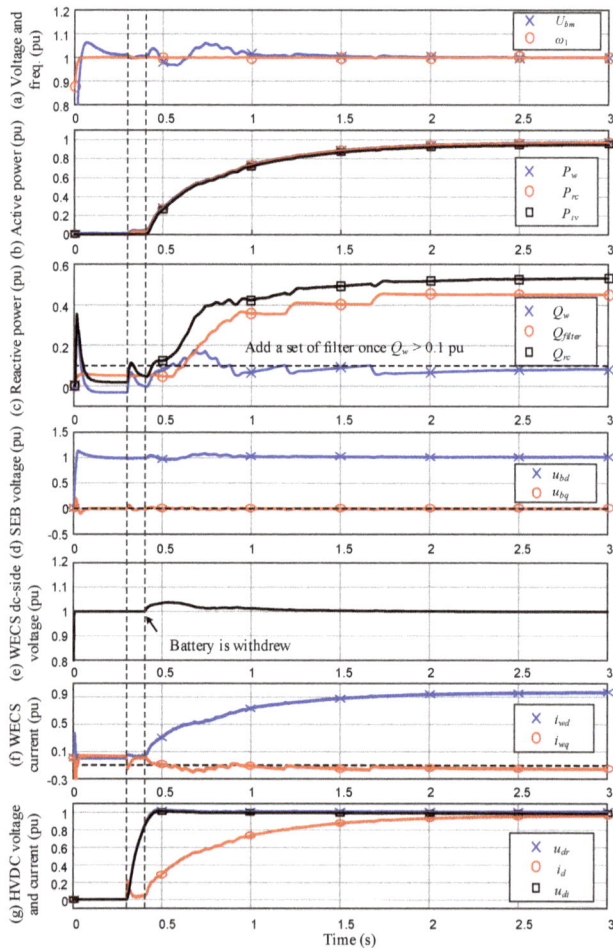

Figure 15. Simulation result of the system startup. (**a**) Sending-end bus voltage and frequency; (**b**) active power; (**c**) reactive power; (**d**) *dq*-axis sending-end bus voltage; (**e**) WECS DC-link voltage; (**f**) WECS output current; (**g**) HVDC DC-link voltage and current.

5.2. Operation under Disturbances

Under disturbances, such as fluctuations of active power and reactive power, the simulation results are shown in Figure 16, where the WF is equivalent to a single WECS with the PI-type Q-f control. In Figure 16, at 0.5 s, the DC-side current of the WECS starts to decrease, which simulates the decrease of the input active power from the front end of the WECS. Figure 16a shows that a small drop of the SEB voltage U_{bm} occurs. Then, the voltage returns to the rated value under the P-V control. At 2.0 s, the DC-side current rises to the original value, and a contrary phenomenon can be observed in Figure 16b. It should be noticed that the AC filters cannot be removed or added due to a time delay. Therefore, the reactive power difference between Q_{rc} and Q_{filter} can be compensated by Q_w. From 4 s to 9 s, several sets of filters are added and removed intentionally, in order to verify the performance of the Q-f control. From Figure 16d, it can be observed that the reactive current i_{wq} automatically changes with the demand for reactive power under the condition that $u_{bq} = 0$, guaranteeing the frequency stability. Accordingly, the reactive power differences between Q_{rc} and Q_{filter} can be automatically compensated by the WECS, as shown in Figure 16c.

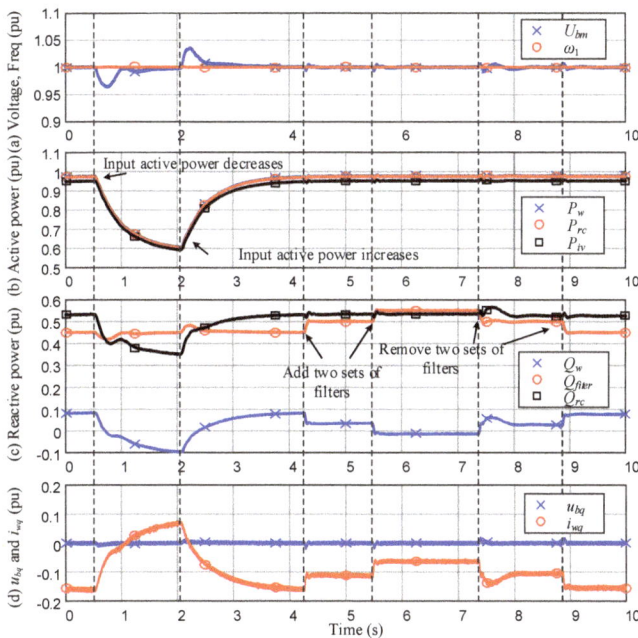

Figure 16. Simulation results under external disturbances, where the WF is represented by a single WECS with the PI-type Q-f control. (**a**) Sending-end bus voltage and frequency; (**b**) active power; (**c**) reactive power; (**d**) dq-axis sending-end bus voltage.

In order to evaluate the performance of the Q-f droop control in a multi-machine WF, the WF is represented by two WECSs with a P-type Q-f droop control. The capabilities of the two WECSs are the same (500 MVA) but the droop coefficients are different (2.0 versus 1.0). The simulation result is shown in Figure 17. The DC-side currents of the WECSs decrease at 0.5 s and then increase at 2.5 s. From 4.5 s to 7.5 s, several sets of AC filters are connected to the SEB gradually. The similar results compared with Figure 16 can be obtained. However, there is a small static error in $u_{bq1,2}$ because of the adopted P-type droop control, as shown in Figure 17d. Moreover, from Figure 17d, it can be also observed that the shared reactive current by WECS$_1$ i_{wq1} is twice that of WECS$_2$ i_{wq2}, due to the relationship between their droop coefficients.

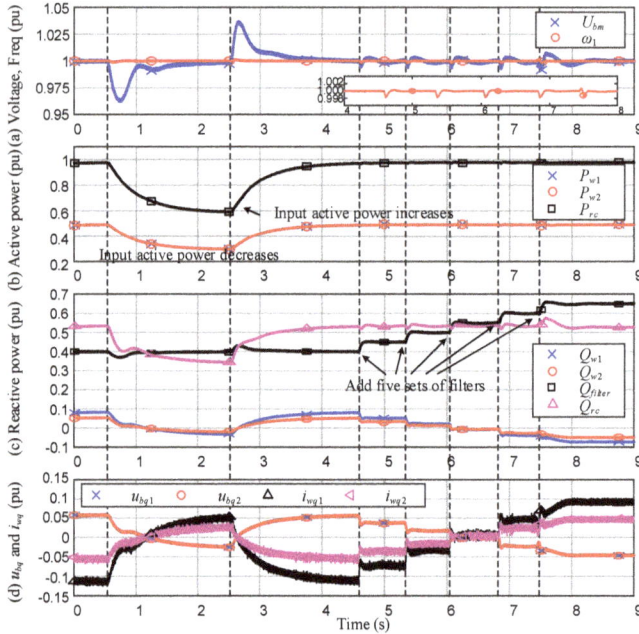

Figure 17. Simulation results under external disturbances, where the WF is represented by two WECSs with the P-type Q-f droop control. (**a**) Sending-end bus voltage and frequency; (**b**) active power; (**c**) reactive power; (**d**) dq-axis sending-end bus voltage. Note that P_{w1}, P_{w2} denote the active powers outputted by both the WECSs, respectively, similar for Q_{w1}, Q_{w2}. Q_{filter} denotes the reactive power generated by the filters.

It should be noted that the dynamic behaviors of the system is affected by both the system parameters and control parameters. The small-signal stability analysis performed in Section 4.3 just illustrates the results of one of the typical operating points. Similar repetitive work can be made to further study the effect of control parameters on the system stability at different operating points. Furthermore, as shown in Figure 9, the phase-plane analysis, together with time-domain state trajectory based on Equations (22) and (23), or Equations (28) and (29) can be employed to evaluate the effects of the system parameters and control parameters on the dynamic behaviors of the system in practical applications.

6. Conclusions

This paper proposed a novel coordinated control scheme for WFs with LCC-HVDC integration. The scheme comprises the Q-f control loop in the WECSs, and the P-V control loop in the LCR. The Q-f control maintains the system frequency and compensates for the reactive power for the LCR automatically, whereas the P-V control maintains the AC bus voltage and realizes the active power balance of the sending-end bus of the HVDC. Thus, the scheme addresses both the voltage and frequency stability, based on the coordination between the WF and the LCR.

The distinguishing features of the scheme can be concluded as follows: (1) there are no commonly used PLLs in the controllers of WECSs, and consequently, the frequency and synchronization stability issues introduced by PLLs can be avoided; (2) the reactive power droop instead of the active power droop is adopted while being applied to achieve synchronization control and reactive power sharing in multi-machine systems, and therefore, the maximum power point tracking of WFs remains unaffected; (3) the scheme can be utilized in more universal scenarios, as long as the core topology is the VSI with LCR connection, such as WFs and photovoltaic power plants with LCC-based rectifier HVDC integration. Our future work will focus on the control and protection algorithms during fault operation, e.g., voltage-dependent current order limits (VDCOLs) for LCC-HVDC and low voltage ride-through (LVRT) for WECS.

Author Contributions: Conceptualization, X.H. and H.G.; Methodology, X.H.; Validation, X.H, H.G. and G.Y.; Writing—Original Draft Preparation, X.H.; Writing—Review & Editing, H.G., G.Y. and X.Z.; Supervision, H.G. and X.Z.; Project Administration, G.Y.

Funding: This work was supported by the National Natural Science Foundation of China (Nos. 61722307, U1510208, and 51711530235).

Conflicts of Interest: The authors declare no conflict of interest.

References

1. Mendoza-Vizcaino, J.; Sumper, A.; Galceran-Arellano, S. PV, Wind and storage integration on small islands for the fulfilment of the 50-50 renewable electricity generation target. *Sustainability* **2017**, *9*, 905. [CrossRef]
2. Korkas, C.D.; Baldi, S.; Michailidis, I.; Kosmatopoulos, E.B. Occupancy-based demand response and thermal comfort optimization in microgrids with renewable energy sources and energy storage. *Appl. Energy* **2016**, *163*, 93–104. [CrossRef]
3. Tavakoli, M.; Shokridehakia, F.; Marzband, M.; Godina, R.; Pouresmaeil, E. A two stage hierarchical control approach for the optimal energy management in commercial building microgrids based on local wind power and PEVs. *Sustain. Cities Soc.* **2018**, *41*, 332–340. [CrossRef]
4. Zhou, H.; Yang, G.; Wang, J. Modeling, analysis, and control for the rectifier of hybrid HVdc systems for DFIG-based wind farms. *IEEE Trans. Energy Convers.* **2011**, *26*, 340–353. [CrossRef]
5. Brenna, M.; Foiadelli, F.; Longo, M.; Zaninelli, D. Improvement of wind energy production through HVDC systems. *Energies* **2017**, *10*, 157. [CrossRef]
6. Smailes, M.; Ng, C.; Mckeever, P.; Shek, J.; Theotokatos, G.; Abusara, M. Hybrid, multi-megawatt HVDC transformer topology comparison for future offshore wind farms. *Energies* **2017**, *10*, 851. [CrossRef]
7. Halawa, E.; James, G.; Shi, X.R.; Sari, N.H.; Nepal, R. The prospect for an Australian-Asian power grid: A critical appraisal. *Energies* **2018**, *11*, 200. [CrossRef]
8. Pena, R.; Clare, J.C.; Asher, G.M. Doubly fed induction generator using back-to-back PWM converters and its application to variable-speed wind-energy generation. *IEE Proc. Electr. Power Appl.* **1996**, *143*, 231–241. [CrossRef]
9. Ma, S.; Geng, H.; Liu, L.; Yang, G.; Pal, B.C. Grid-Synchronization stability improvement of large scale wind farm during severe grid fault. *IEEE Trans. Power Syst.* **2018**, *33*, 216–226. [CrossRef]
10. Zhou, H.; Yang, G.; Wang, J.; Geng, H. Control of a hybrid high-voltage DC connection for large doubly fed induction generator-based wind farms. *IET Renew. Power Gener.* **2011**, *5*, 36–47. [CrossRef]
11. Zhou, H.; Yang, G.; Geng, H. Grid integration of DFIG-based offshore wind farms with hybrid HVDC connection. In Proceedings of the 2008 International Conference on Electrical Machines and Systems, Wuhan, China, 17–20 October 2008.
12. Bozhko, S.V.; Blasco-Gimenz, R.; Li, R.; Clare, J.C.; Asher, G.M. Control of offshore DFIG-based wind farm grid with line-commutated HVDC connection. *IEEE Trans. Energy Convers.* **2007**, *22*, 71–78. [CrossRef]
13. Bozhko, S.V.; Asher, G.; Li, R.; Clare, J.; Yao, L. Large offshore DFIG-based wind farm with line-commutated HVDC connection to the main grid: Engineering studies. *IEEE Trans. Energy Convers.* **2008**, *23*, 119–127. [CrossRef]
14. Xiang, D.; Ran, L.; Bumby, J.R.; Tavner, P.J.; Yang, S. Coordinated control of an HVDC link and doubly fed induction generators in a large offshore wind farm. *IEEE Trans. Power Deliv.* **2006**, *21*, 463–471. [CrossRef]
15. Li, R.; Bozhko, S.; Asher, G. Frequency control design for offshore wind farm grid with LCC-HVDC link connection. *IEEE Trans. Power Electron.* **2008**, *23*, 1085–1092. [CrossRef]

16. Zhang, M.; Yuan, X.; Hu, J.; Wang, S.; Ma, S.; He, Q.; Yi, J. Wind power transmission through LCC-HVDC with wind turbine inertial and primary frequency supports. In Proceedings of the 2015 IEEE Power & Energy Society General Meeting, Denver, CO, USA, 26–30 July 2015.

17. Cardiel-Álvarez, M.Á.; Rodriguez-Amenedo, J.L.; Arnaltes, S.; Montilla-DJesus, M.E. Modeling and control of LCC rectifiers for offshore wind farms connected by HVDC links. *IEEE Trans. Energy Convers.* **2017**, *32*, 1284–1296. [CrossRef]

18. Yin, H.; Fan, L.; Miao, Z. Coordination between DFIG-based wind farm and LCC-HVDC transmission considering limiting factors. In Proceedings of the IEEE 2011 EnergyTech, Cleveland, OH, USA, 25–26 May 2011.

19. Yin, H.; Fan, L. Modeling and control of DFIG-based large offshore wind farm with HVDC-link integration. In Proceedings of the 41st North American Power Symposium, Starkville, MS, USA, 4–6 October 2009.

20. Blasco-Gimenez, R.; Añó-Villalba, S.; Rodríguez-D'Derlée, J.; Morant, F.; Bernal-Perez, S. Distributed voltage and frequency control of offshore wind farms connected with a diode-based HVdc link. *IEEE Trans. Power Electron.* **2010**, *25*, 3095–3105. [CrossRef]

21. Yu, L.; Li, R.; Xu, L. Distributed PLL-based control of offshore wind turbine connected with diode-rectifier based HVDC systems. *IEEE Trans. Power Deliv.* **2018**, *33*, 1328–1336. [CrossRef]

22. Cardiel-Álvarez, M.Á.; Arnaltes, S.; Rodriguez-Amenedo, J.L.; Nami, A. Decentralized control of offshore wind farms connected to diode-based HVDC links. *IEEE Trans. Energy Convers.* **2018**, *33*, 1233–1241. [CrossRef]

23. Xi, X.; Geng, H.; Yang, G. Enhanced model of the doubly fed induction generator-based wind farm for small-signal stability studies of weak power system. *IET Renew. Power Gener.* **2014**, *8*, 765–774. [CrossRef]

24. Zhang, Y.; Ooi, B.T. Stand-Alone doubly-fed induction generators (DFIGs) with autonomous frequency control. *IEEE Trans. Power Deliv.* **2013**, *28*, 752–760. [CrossRef]

25. Pena, R.; Clare, J.C.; Asher, G.M. A doubly fed induction generator using back-to-back PWM converters supplying an isolated load from a variable speed wind turbine. *IEE Proc. Electr. Power Appl.* **1996**, *143*, 380–387. [CrossRef]

26. Abdoune, F.; Aouzellag, D.; Ghedamsi, K. Terminal voltage build-up and control of a DFIG based stand-alone wind energy conversion system. *Renew. Energy* **2016**, *97*, 468–480. [CrossRef]

27. Li, D.; Shen, Q.; Liu, Z.; Wang, H.; Ding, M.; Liu, F. Auto-disturbance rejection control for the stator voltage in a stand-alone DFIG-based wind energy conversion system. In Proceedings of the 2016 35th Chinese Control Conference (CCC), Chengdu, China, 27–29 July 2016.

28. Fazeli, M.; Asher, G.; Klumpner, C.; Yao, L. Novel integration of DFIG-based wind generators within microgrids. *IEEE Trans. Energy Convers.* **2011**, *26*, 840–850. [CrossRef]

29. Fazeli, M.; Bozhko, S.V.; Asher, G.M.; Yao, L. Voltage and frequency control of offshore DFIG-based wind farms with line commutated HVDC connection. In Proceedings of the 2008 4th IET Conference on Power Electronics, Machines and Drives, York, UK, 2–4 April 2008.

30. Huang, L.; Xin, H.; Zhang, L.; Wang, Z.; Wu, K.; Wang, H. Synchronization and frequency regulation of DFIG-based wind turbine generators with synchronized control. *IEEE Trans. Energy Convers.* **2017**, *32*, 1251–1262. [CrossRef]

31. Zhong, Q.-C.; Nguyen, P.-L.; Ma, Z.; Sheng, W. Self-synchronized synchronverters: Inverters without a dedicated synchronization unit. *IEEE Trans. Power Electron.* **2014**, *29*, 617–630. [CrossRef]

32. Kalman, R.E. Phase-plane analysis of automatic control systems with nonlinear gain elements. *Trans. AIEE* **1955**, *73*, 383–390. [CrossRef]

33. Atighechi, H.; Chiniforoosh, S.; Jatskevich, J.; Davoudi, A.; Martinez, J.A.; Faruque, M.O.; Sood, V.; Saeedifard, M.; Cano, J.M.; Mahseredjian, J.; et al. Dynamic average-value modeling of CIGRE HVDC benchmark system. *IEEE Trans. Power Deliv.* **2014**, *29*, 2046–2054. [CrossRef]

MDPI

St. Alban-Anlage 66

4052 Basel

Switzerland

Tel. +41 61 683 77 34

Fax +41 61 302 89 18

www.mdpi.com

Energies Editorial Office

E-mail: energies@mdpi.com

www.mdpi.com/journal/energies

www.ingramcontent.com/pod-product-compliance
Lightning Source LLC
Chambersburg PA
CBHW051853210326
41597CB00033B/5887